SpringerBriefs in Electrical and Computer Engineering

SpringerBriefs in Computational Electromagnetics

Series Editors

K. J. Vinoy, Indian Institute of Science, Professor, Electrical Communication Engineering, Indian Institute of Science, Bangalore, Karnataka, India

Rakesh Mohan Jha (Late), Centre for Electromagnetics, CSIR-National Aerospace Laboratories, Bangalore, Karnataka, India

W0080358

The phenomenon of electric and magnetic field vector (wave) propagation through the free-space, or any other medium is considered within the ambit of electromagnetics. The media themselves, in general, could be of diverse type, such as linear/non-linear, isotropic/non-isotropic, homogeneous/inhomogeneous, reciprocal/non-reciprocal, etc. Such electromagnetic wave propagation problems are formulated with the set of Maxwell's equations. Computational Electromagnetics endeavors to provide the solution to the Maxwell's equations for a given formulation. It is often difficult to find closed forms solutions to the Maxwell's equation formulations. The advent of computers, and particularly the initial developments of efficient coding for numerical analysis, encouraged the development of numerical electromagnetics. A second motivation came from the interaction of the electromagnetic wave with the matter. This could be visualized as scattering bodies, which required incorporation of the phenomena of reflection, refraction, diffraction and polarization. The finite/large nature of the scatterer required that problem of electromagnetics is considered with respect to the operation wavelength leading to the classification of low-frequency, high-frequency and resonance region problems. This also inspired various asymptotic and grid-based finite-method techniques, for solving specific electromagnetic problems. Surface modeling and ray tracing are also considered for such electromagnetic problems. Further, design optimization towards hardware realization have led to the recourse to various soft computing algorithms. Computational Electromagnetics is deemed to encompass the numerical electromagnetics along with all other above developments. With the wide availability of massively parallel high performance parallel computing platforms, new possibilities have emerged for reducing the computation time and developing macro models that can even be employed for several practical multi-physics scenarios. Both volume and surface discretization methods have been given a new boost, and several acceleration techniques including GPU based computation, learning based approaches, and model order reduction have been attempted. Limitations of generating meshes and modifying these for parametric estimation have been addressed by statistical approaches and smart solvers. Many nature-inspired algorithms and other soft computing approaches have been employed for electromagnetic synthesis problems. One of the recent additions is Game Theoretic optimization.

Finally, the emergence of Computational Electromagnetics has been motivated by myriad applications. These diverse application include but are not restricted to those in Electronics and Communication, Wireless Propagation, Computer Hardware, Aerospace Engineering, Biomedical Engineering, Radio-astronomy, Terahertz Technology, Photonics, etc for modelling of devices, components, systems and even large structures. Some of the well-known applications are in Analysis and design of radio frequency (RF) circuit, antennas and systems, Analysis of antenna on structures, radar imaging, radar cross section (RCS) computation and reduction, and analysis of electromagnetic wave-matter interactions at discrete, random and periodic geometries including metamaterials. Authors are encouraged to submit original research work in the area of Computational Electromagnetics. The content could be either theoretical development, or specific to particular applications. This Series also encourages state-of-the-art reviews and easy to comprehend tutorials.

More information about this subseries at http://www.springer.com/series/13885

P. S. Mohammed Yazeen ·
C. V. Vinisha · S. Vandana ·
M. Suprava · Raveendranath U. Nair

Broadbanding Techniques for Radomes

 Springer

P. S. Mohammed Yazeen
Centre for Electromagnetics (CEM)
CSIR-National Aerospace Laboratories
Bengaluru, Karnataka, India

C. V. Vinisha
Centre for Electromagnetics (CEM)
CSIR-National Aerospace Laboratories
Bengaluru, Karnataka, India

S. Vandana
Centre for Electromagnetics (CEM)
CSIR-National Aerospace Laboratories
Bengaluru, Karnataka, India

M. Suprava
Centre for Electromagnetics (CEM)
CSIR-National Aerospace Laboratories
Bengaluru, Karnataka, India

Raveendranath U. Nair
Centre for Electromagnetics (CEM)
CSIR-National Aerospace Laboratories
Bengaluru, Karnataka, India

ISSN 2191-8112   ISSN 2191-8120 (electronic)
SpringerBriefs in Electrical and Computer Engineering
ISSN 2365-6239   ISSN 2365-6247 (electronic)
SpringerBriefs in Computational Electromagnetics
ISBN 978-981-33-4129-6   ISBN 978-981-33-4130-2 (eBook)
https://doi.org/10.1007/978-981-33-4130-2

This Springer imprint is published by the registered company Springer Nature Singapore Pte Ltd.
The registered company address is: 152 Beach Road, #21-01/04 Gateway East, Singapore 189721, Singapore

To

Late Dr. Rakesh Mohan Jha

Founder Scientist
of
Centre for Electromagnetics
CSIR-NAL Bangalore, India

Preface

Broadbanding of radomes has been widely used for the enhancement of transmission efficiency of the radomes. Major goals while broadbanding are to obtain high transmission efficiency, low cross-polarization levels, and low boresight error characteristics over the entire broadband with requisite modifications of radome wall configuration. Many techniques are employed for modifying the radome wall configurations for enhancing its EM performance.

Modification of the radome wall configuration can be accomplished by using different techniques like (i) inclusion of metallic wiregrids in the radome walls, (ii) inclusion of metallic strip gratings in the radome layers, (iii) inclusion of FSS-based structures in between the radome layers, and (iv) use of inhomogeneous dielectric structures as radome wall.

This book presents a detailed formulation for the broadbanding of radomes using various techniques. This book is organized into six chapters: Chapter 1 deals with the introduction of radomes, Chap. 2 deals with the EM performance characteristics of radomes, Chap. 3 includes the broadbanding techniques based on the metallic wiregrids, Chap. 4 is about the broadbanding techniques based on the metallic strip gratings, and in Chap. 5, broadbanding techniques based on the FSS structures are discussed. The broadbanding of radomes based on inhomogenous planar layers is described in Chap. 6, and finally, it concludes with the summary of various broadbanding techniques discussed in the brief.

Bengaluru, India

P. S. Mohammed Yazeen
C. V. Vinisha
S. Vandana
M. Suprava
Raveendranath U. Nair

Acknowledgements

We would like to thank Mr. Jitendra J Jadhav, Director, CSIR-National Aerospace Laboratories, Bangalore, for giving permission to write this Springer Brief.

Further, we record our thanks to colleagues Dr. Hema Singh, Dr. Shiv Narayan, Dr. Balamati Choudhury, Ms. Vineeetha Joy for their suggestions and cooperations. We express our sincere thanks to Project Staff of Centre for Electromagnetics Ms. Karthika S. Nair, Ms. Maumita Dutta, and Ms. Mahima P. for their consistent support during the preparation of the technical contents of this book.

We express our gratitude to Dr. Suvira Srivastav, Associate Director, Springer (India) Private Limited, for giving us opportunity to write this book. Nevertheless, it would have not been possible to bring out this book within a short span of time without consistent support and suggestions of Ms. Swati Meherishi, Editorial Director, Applied Sciences and Engineering. We highly appreciate the effort made by Ms. Kamiya Khatter, Associate Editor, Applied Sciences, and Ms. Aparajita Singh, Editorial Assistant, Applied Sciences, to accomplish this book.

About This Book

Modern airborne radomes have very stringent electromagnetic (EM) performance requirements. Hence novel techniques have to be used in the EM design of airborne radomes to meet these requirements. Broadbanding of radomes has been widely used for the enhancement of transmission efficiency of the radomes. This brief deals with various techniques employed for enhancing the EM performance parameters by modifying the radome wall configurations.

Modification of the radome wall configuration has been accomplished by using different techniques such as inclusion of metallic wiregrids/meshes in the radome walls, inclusion of metallic stripgratings in the radome layers, inclusion of FSS based structures in between the radome layers and the use of inhomogeneous dielectric structures as radome wall. This book presents a detailed formulation for the broadbanding of radomes using these techniques.

Contents

About the Authors

P. S. Mohammed Yazeen is currently an Associate Staff Engineer (RF) with Samsung R&D Institute, Bangalore, India. He previously worked as a project scientist at Centre for Electromagnetics (CEM), CSIR-National Aerospace Laboratories (CSIR-NAL), Bangalore, India. He obtained his Master's degree in Electronics with specialization in Wireless Technology from Cochin University of Science and Technology (CUSAT), Kerala, India in 2014. His research interests are design and analysis of streamlined airborne radomes, wireless communication, digital beamformers, EM material characterization, RF circuits, RF driver development, etc.

C. V. Vinisha is currently working as Research Scholar in Cochin University of Science and Technology (CUSAT), Kerala, India. She previously worked as a project scientist at Centre for Electromagnetics (CEM), CSIR-National Aerospace Laboratories (CSIR-NAL), Bangalore, India. She received her M.Tech. in Electronics with specialization in Microwave and Radar Engineering from CUSAT, India in 2015. Her research interests are in computational electromagnetics, radomes, and antenna arrays.

S. Vandana worked as a project trainee at Centre for Electromagnetics (CEM), CSIR-National Aerospace Laboratories (CSIR-NAL), Bangalore, India. She obtained her B.Tech. in Electronics and Communication Engineering from College of Engineering Kalloopara, Kerala, India in 2010 and M.Tech. in Electronics with specialization in Digital Electronics from CUSAT, India in 2013. Her research interests are design and analysis of radomes, signal processing and EM material characterization techniques.

M. Suprava worked as project engineer at Centre for Electromagnetics (CEM), CSIR-National Aerospace Laboratories (CSIR-NAL), Bangalore, India. She obtained her B.Tech. in Electronics and Communication Engineering from Vignan Institute of Technology and Management, India in 2010. Her research interests are in design and analysis of radomes.

Dr. Raveendranath U. Nair is currently Senior Principal Scientist and Head, Centre for Electromagnetics (CEM), CSIR-National Aerospace Laboratories (CSIR-NAL), Bangalore, India. He received his M.Sc (Physics) and Ph.D. in Physics (Microwave Electronics) from the School of Pure and Applied Physics, Mahatma Gandhi University, Kerala, India, in 1989 and 1997, respectively. He has authored/co-authored over 150 research publications including peer reviewed journal papers, symposium papers and technical reports. Dr. Nair received the CSIR-NAL Excellence in Research Award (2007-2008) for his contributions to the EM design of variable thickness airborne radomes. Dr. Nair is also a Professor of the Academy of Scientific and Innovative Research (AcSIR), New Delhi.

Abbreviations

BSE	Boresight Error
DSL-FSS	Double Square Loop-Frequency Selective Surface
EM	Electromagnetics
FSS	Frequency Selective Surfaces
IPD	Insertion Phase Delay
IPL	Inhomogeneous Planar Layer
JC-FSS	Jerusalem Cross Frequency Selective Surface
SLL	Sidelobe Level
SSL	Single Square Loop
SSL-FSS	Single Square Loop—Frequency Selective Surface

Symbols

ε	Dielectric constant
ε_m	Relative dielectric constant of mixture
ε^*	Complex permittivity
ε_c	Dielectric constant of the core
ε_F	Relative dielectric constant of reinforcement fibers
ε_R	Relative dielectric constant of resin
ε_r	Relative dielectric constant
ε_{rm}	Maximum value of dielectric constant
λ	Wavelength of incident wave
θ	Incidence angle
Φ	Electrical length
$\angle T$	Phase angle associated with the voltage transmission coefficient
A_i, B_i, C_i, and D_i	A, B, C, D parameters of ith dielectric layer
A_{FSS}, B_{FSS}, C_{FSS} and D_{FSS}	Elements of the matrix representing aperture type SSL-FSS
A_{MS}, B_{MS}, C_{MS} and D_{MS}	Elements of the matrix representing metallic strip grating
A_{WG}, B_{WG}, C_{WG} and D_{WG}	Elements of the matrix representing the wiregrid
B	Capacitive susceptance
B_{TE}	Capacitive susceptance of TE mode of incidence
$C_{1\pm}$	Correction term for shunt susceptance
d	Thickness of the dielectric layer
D	Diameter of the wire
d_{FSS}	Thickness of SSL-FSS
d_m	Radome wall thickness
F	Correction term for Jerusalem cross frequency selective surface
g	Gap length
G	Correction term for shunt susceptance

p	Periodicity
P	Pitch of the wiregrid
P_{rf}	Power reflection coefficient
P_{tr}	Power transmission coefficient
R	Voltage reflection coefficient
T	Voltage transmission coefficient
t_s	Thickness of skin layer
t_{ms}	Thickness of metallic strip grating
t_c	Thickness of core
$\tan\delta_e$	Loss tangent
V_F	Volume of the reinforcement fibers
V_R	Volume of the resin
w	Width of SSL element
X	Inductive reactance
X_r	Reactance of the SSL-FSS
X_{TE}	Inductive reactance for TE mode of incidence
X_{TM}	Inductive reactance of TM mode of incidence
Z_0	Characteristic impedance of free space
Z_c	Characteristic impedance of core
Z_G	Characteristic impedance of wiregrid
Z_s	Characteristic impedance of skin
Z_{FSS}	Input impedance of FSS

List of Figures

List of Tables

Chapter 1
Introduction

Radome is a structure used to shield and protect the radar antenna from damages and adverse environmental conditions in ground-based, shipboard, and airborne applications.

Antennas used in airborne applications are affected by high aerodynamic stress caused due to the high flight speed of aircraft. Airborne radomes aid in the protection of these antennas with minimal degradations of antenna radiation characteristics.

Modern airborne radomes have very stringent electromagnetic (EM) performance requirements. Hence, novel techniques have to be used in the EM design of airborne radomes to meet these requirements.

In conventional radome wall configurations, achieved EM performance characteristics such as high transmission efficiency, low cross-polarization levels, and low boresight error, are not satisfactory over a wide range of incidence angles. Further enhancement of radome performance parameters can be obtained by two methods: (i) modification of the thickness profile of the radome wall (tapering of radome wall) and (ii) modification of the radome wall configuration with inclusions.

In the first technique, continuous tapering or graded tapering of the wall is employed depending on the EM performance requirements. The limitation of this method is that the EM performance parameters can be improved only at the given design frequency and incidence angle. Hence, this becomes ineffective for radome broadbanding applications.

Numerous techniques have been developed for improving the EM performance of radomes (Walton 1966; Frenkel 2001; Nair et al. 2013a, b). Most of these techniques were based on (a) metallic wiregrids (b) metallic wire meshes, (c) resonant/semi-resonant inclusions and (d) frequency selective surfaces (FSS). In few works reported previously (Richmond 1964) inhomogeneous planar layers (IPL) had been incorporated for the design of radomes. IPL **is** planar structures with varying dielectric parameters (dielectric constant and loss tangent) across the thickness of the wall.

For the design of large airborne nosecone radomes, usage of monolithic half-wave wall configurations is avoided owing to the increase in weight and their very

© The Author(s), under exclusive license to Springer Nature Singapore Pte Ltd. 2020   1
P. S. Mohammed Yazeen et al., *Broadbanding Techniques for Radomes*,
SpringerBriefs in Computational Electromagnetics,
https://doi.org/10.1007/978-981-33-4130-2_1

narrow bandwidth. Multilayer configurations like A-sandwich or C-sandwich are also avoided in spite of having high strength-to-weight ratio as their EM performance parameters over wide frequency range are not sufficient to meet the stringent performance requirements of the modern fire control radar systems. For the enhancement of EM performance characteristics, the following techniques can be adopted to modify the monolithic and multilayered radome wall configurations.

(a) Inclusion of metallic grids/strip gratings in the radome wall.
(b) Inclusion of frequency selective surfaces (FSS) in the radome wall.
(c) Use of inhomogeneous dielectric structures as radome wall.

The metallic inclusions used for broadbanding purpose can deteriorate the EM performance of the radome as these structures are sensitive to polarization and angle of incidence. Thus, the design parameters for the metallic structures have to be appropriately selected to obtain the desired performance.

In this brief, several techniques for broadbanding of radomes have been demonstrated. The analysis of the modified structures is done using the equivalent transmission line model. A detailed chapter-wise explanation for the design aspects and performance analysis of the modified radome wall configurations has been discussed in this book.

References

Frenkel A (2001) Thick metal dielectric radome. IET Electron Lett 38:1374–1375
Nair RU, Madhumitha J, Jha RM (2013a) Broadband EM performance characteristics of single square loop FSS embedded monolithic radome. Int J Antennas Propag:1–8
Nair RU, Madhumitha J, Jha RM (2013b) Application of metallic strip gratings for enhancement of electromagnetic performance of A-sandwich radome. Defense Sci J 63:508–514
Richmond JH (1964) On the design of one layer and three-layer tunable radomes.The Ohio State University, Department of Electrical Engineering, Antenna Laboratory, Report 1751-4 (ASTIA Document No. AD 433839), 15 Jan 1964
Walton JD (1966) Techniques for airborne radome design, AFAL report, no. 45433. Wright-Patterson AFB, pp 105–108

Chapter 2
EM Performance Characteristics of Radomes

The interaction of EM waves with the radome wall must be considered based on first-order effects (such as amplitude and phase distortions) during the EM design of the radome wall. The estimations of the basic performance parameters primarily depend on the EM material parameters of constituent layers of radome wall (dielectric constant, electric loss tangent, etc.), thickness of constituent layers, dimensions and periodicities of unit cell elements of embedded structures, operating frequency and polarization of the antenna system, and the range of incidence angles by the antenna beam during scanning. Incidence angles can be evaluated based on the location of antenna system within the radome and antenna-gimbal rotational offsets (Nair and Rakesh 2014).

2.1 Radome Wall Configurations

Organic materials are mainly used for the fabrication of civilian and military aircraft, ground vehicles, and fixed terrestrial radomes. Typical design techniques for radome wall construction are based on either solid monolithic or sandwich design. Resins incorporating reinforcement of fibers are generally used for monolithic design whereas relatively high- and low-density (dielectric constant) materials are alternatively used for the sandwich design. Equation for the relative dielectric constant of the mixture is given as (Kozakoff 2010).

© The Author(s), under exclusive license to Springer Nature Singapore Pte Ltd. 2020   3
P. S. Mohammed Yazeen et al., *Broadbanding Techniques for Radomes*,
SpringerBriefs in Computational Electromagnetics,
https://doi.org/10.1007/978-981-33-4130-2_2

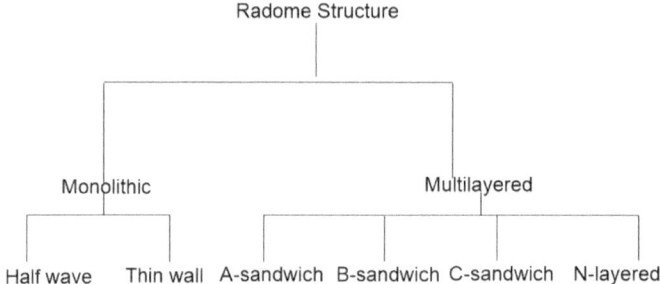

Fig. 2.1 Classification of radome wall configurations

$$\epsilon_m = \frac{V_R \log \epsilon_R + V_F \log \epsilon_F}{V_R + V_F} \tag{2.1}$$

where

ϵ_m relative dielectric constant of mixture.
ϵ_R relative dielectric constant of resin.
ϵ_F relative dielectric constant of reinforcement fibers.
V_F volume of the reinforcement fibers.
V_R volume of the resin.

Depending on the number of layers, the radome wall configurations are classified into several types. The radome wall classifications are given in Fig. 2.1.

2.1.1 Monolithic Half-Wave Wall Configuration

The monolithic half-wave wall configuration consists of a single dielectric layer. Its thickness is in multiples of half wavelength in the dielectric. This configuration offers minimum internal reflections.

2.1.2 A-sandwich Wall Configuration

A-sandwich radome wall consists of two skins layers separated by a dielectric core. The thickness of the core is much higher than that of skins, whereas the dielectric constant of the core is less than that of skin. The structural rigidity and bandwidth are greater for A-sandwich radome wall configuration as compared to the solid-wall radome.

2.1.3 B-sandwich Wall Configuration

This wall configuration is made of a thick, high dielectric solid core placed between two skins having low dielectric constant. For perpendicular as well as parallel polarizations, equal transmission characteristics are obtained in the case of B-sandwich radome wall configurations. The skin layer with low dielectric constant acts as a quarter-wave transformer.

2.1.4 C-sandwich Wall Configuration

C-sandwich consists of five layers, i.e., inner and outer skins, a middle layer, and two cores. The dielectric constant of the core is lower than that of the skins and middle layer. This structure can be considered as two A-sandwiches placed back to back. The C-sandwich configuration has excellent structural rigidity as compared to the conventional radomes. As compared to the A-sandwich radome wall configurations, C-sandwich has better transmission efficiency and bandwidth. This configuration enables further cancelation of residual reflections that arises from the individual A-sandwiches.

2.1.5 Multilayered Wall Configuration

The multilayer consists of different thin skins and cores of varying dielectric constants. The multilayered wall configurations have excellent structural rigidity and broadband transmission characteristics for all incidence angles. These types of radome wall configurations offer good electrical performance over wide bandwidth and better resistance to thermal shock and rainfall.

2.2 Equivalent Transmission Line Method

The multilayered radome wall can be considered as an equivalent transmission line with each section corresponding to individual dielectric layers. In this method, each dielectric layer is represented by a section of a transmission line (Cary et al. 1983). As compared to free space, different dielectric layers can be considered as low impedance lines connected end to end. Variation of the characteristic impendence from free space to the radome wall is a major cause for incident wave power reflection on the structure. For EM analysis, the whole configuration can be represented by a single matrix obtained by the multiplication of matrices corresponding to the individual dielectric layers.

The matrix describing the single layer is given as

$$
\begin{bmatrix}
\cos \Phi & j\frac{Z}{Z_0} \sin \Phi \\
j\frac{Z_0}{Z} \sin \Phi & \cos \Phi
\end{bmatrix}
\tag{2.2}
$$

where the electrical length is

$$
\phi = \frac{2\pi d \sqrt{\varepsilon - \sin^2 \theta}}{\lambda}
\tag{2.3}
$$

Here

d thickness of the layer.
λ wavelength of incident wave.
ε dielectric constant.
θ incidence angle.

Here, Z/Z_0 is the ratio of the impedance in the medium to that in the free space.

$$
\frac{Z}{Z_0} = \frac{\sqrt{\varepsilon - \sin^2 \theta}}{\varepsilon \cos \theta} \quad \text{for parallel polarization}
\tag{2.4}
$$

$$
\frac{Z}{Z_0} = \frac{\cos \theta}{\sqrt{\varepsilon - \sin^2 \theta}} \quad \text{for perpendicular polarization}
\tag{2.5}
$$

The multilayered radome wall configuration is represented in matrix form as,

$$
\begin{bmatrix}
A & B \\
C & D
\end{bmatrix} = [1][2]\ldots[n]\,\text{layers}
\tag{2.6}
$$

where $[1]$, $[2]$ … and $[n]$ represents the matrices of individual layers.
    The EM performance parameters are given by.
,

$$
\text{Voltage transmission coefficient, } T = \frac{2}{A + B + C + D}
\tag{2.7}
$$

$$
\text{Voltage reflection coefficient, } R = \frac{A + B - C - D}{A + B + C + D}
\tag{2.8}
$$

$$
\text{Insertion phase delay (IPD), } \Phi = -\angle T - \frac{2\pi d}{\lambda} \cos \theta
\tag{2.9}
$$

where $\angle T$ is the phase angle associated with the complex term T.

References

Kozakoff DJ (2010) Analysis of radome enclosed antennas. Artech House, Norwood. ISBN 1596934417, 294 p

Nair RU, Rakesh MJ (2014) Electromagnetic design and performance analysis of airborne radomes: trends and perspectives. IEEE Antennas Propag Mag 56(4)

Chapter 3
Broadbanding Techniques Based on Metallic Wiregrids

Substantial enhancement in EM performance over a broadband is one of the manda-tory requirements for modern airborne radar systems. Different methods for broad-banding of radomes have been observed in the literatures (Harder and Gaurino 1964; Richmond 1964; Tricoles 1958). Techniques based on embedded metallic wire grids in the radome panels also have been widely used for improving the EM perfor-mance of radomes (Cary 1983; Kay 1958; Miller and Brown 1957; Robinson 1958; Worthington 1958; Walton 1966).

Radome structures having high strength-to-weight ratio, multilayered wall config-uration is chosen over monolithic half-wave wall for manufacturing streamlined airborne radomes. C-sandwich configuration offers better power transmission characteristics than A-sandwich for a range of incidence angles.

For analyzing the EM performance using wiregrids, the power transmission char-acteristics of an optimized core C-sandwich configuration over a broadband of 2–18 GHz is considered.

The EM performance parameters can be improved up to a certain extent by the addition of metallic structures in the skin and core layers of the radome. Thus, design parameters of the wire grids should be optimized to provide enhanced EM performance for normal incidence and high incidence angles.

3.1 EM Design Aspects of Metallic Wiregrids Embedded C-sandwich Radome Panel

C-sandwich wall consists of two skins, two cores and a middle layer, with each core being embedded between a skin and the middle layer. In the current literature, the skin and the middle layer is considered as glass composite (relative permittivity, $\varepsilon_r = 4.0$; electric loss tangent, $\tan\delta_e = 0.015$) with two identical foam cores ($\varepsilon_r = 1.15$; $\tan\delta_e = 0.002$). For C-sandwich, wall configuration thickness of the center layer is

© The Author(s), under exclusive license to Springer Nature Singapore Pte Ltd. 2020
P. S. Mohammed Yazeen et al., *Broadbanding Techniques for Radomes*,
SpringerBriefs in Computational Electromagnetics,
https://doi.org/10.1007/978-981-33-4130-2_3

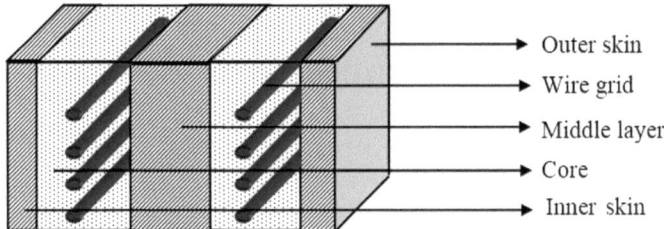

Fig. 3.1 Cross section of C-sandwich radome panel with metallic wire grid embedded in the mid-plane of the core

Table 3.1 Design parameters of metallic wire grids

Angle of incidence (°)	Polarization	Optimum diameter (mm)	Optimum pitch (mm)
0	–	1.4	95.2
45	Perpendicular	1.4	79.6
60	Perpendicular	0.9	65.2

kept at 1.5 mm, and an equal thickness of 0.75 mm is given to the inner and outer skin layers. An optimum thickness of 5.44 mm is assigned to each core to achieve maximum power transmission across the frequency band of 2–18 GHz.

As shown in Fig. 3.1, planar array of wire grids made up of parallel wires having circular cross sections is symmetrically embedded in the center plane of each core. Along the length of the wire material characteristics of the wire grid are the same. Besides the polarization of the incident wave, the pitch and diameter of the wires influences the EM performance parameters. When the electric field of the incident wave is polarized in the direction of the wires, the effect of the wiregrid will be maximum.

To obtain superior EM performance, the capacitive susceptance of the dielectric material should match with the inductive susceptance of the wire grid, which is obtained by optimizing the design parameters of wiregrids. The optimized design parameters of the wire grid are calculated at 10 GHz which is given in Table 3.1.

3.2 EM Performance Analysis of Metallic Wiregrids Embedded C-sandwich Radome Panel

The equivalent transmission line method is used for the estimation of EM performance parameters as explained in Chap. 2. An equivalent transmission line model can be derived from the metallic wiregrids included C-sandwich radome with different sections corresponding to skin, core, and wire grids. Discontinuity in the line caused by variation of the characteristic impendence from free space to the C-sandwich radome wall configuration is a major cause for incident wave power reflection on the structure. A matrix consisting of A_i, B_i, C_i, and D_i parameters represents ith dielectric layer. Hence, the whole configuration can be represented by a single matrix obtained by the multiplication of matrices corresponding to individual layers.

Free space characteristic impedance is given as Z_0, and the characteristic impedances of the skin, core, and wire grid are represented by Z_s, Z_c, and ZG, respectively. Electrical length of each layer is represented as Φ, which is a function of the angle of incidence (θ), complex permittivity $(\varepsilon*)$ of dielectric layer, thickness of the dielectric layer (d), and wavelength of incident wave (λ).

The electrical length (each layer) can be represented as

$$\Phi = \frac{2\pi d\sqrt{\varepsilon* - \sin^2\theta}}{\lambda} \tag{3.1}$$

The matrix representation of each layer of C-sandwich wall is represented as: The outer skin:

$$\begin{bmatrix} A_1 & B_1 \\ C_1 & D_1 \end{bmatrix} = \begin{bmatrix} \cos\Phi_1 & j\frac{Z_s}{Z_0}\sin\Phi_1 \\ j\frac{Z_0}{Z_s}\sin\Phi_1 & \cos\Phi_1 \end{bmatrix} \tag{3.2}$$

In the current wall configuration, the wire grid is placed in the center of each core. Hence, each core can be divided into two similar sections with wire grid embedded in between them. Then, each section of the outer core can be represented by

$$\begin{bmatrix} A_2 & B_2 \\ C_2 & D_2 \end{bmatrix} = \begin{bmatrix} \cos\Phi_2 & j\frac{Z_c}{Z_0}\sin\Phi_2 \\ j\frac{Z_0}{Z_c}\sin\Phi_2 & \cos\Phi_2 \end{bmatrix} \tag{3.3}$$

Let A_{WG}, B_{WG}, C_{WG}, and D_{WG} be the elements of the matrix representing wire grid. Then,

$$\begin{bmatrix} A_{WG} & B_{WG} \\ C_{WG} & D_{WG} \end{bmatrix} = \begin{bmatrix} 1 & 0 \\ jB_G & 1 \end{bmatrix} \tag{3.4}$$

Here, B_G represents the shunt susceptance of the wire grid. For perpendicular polarization,,

$$B_G = \frac{-1}{\left(\frac{P}{\lambda}\right)\sqrt{\varepsilon_c - \sin^2\theta}\left[\log_e\left(\frac{P}{\pi D}\right) + 0.6\left(\frac{P}{\lambda}\right)^2(\varepsilon_c + 2\sin^2\theta)\right]} \tag{3.5}$$

P, D, ε_c and λ represent pitch of the wire grid, wire diameter, dielectric constant, and wavelength, respectively.

The middle layer:

$$\begin{bmatrix} A_3 & B_3 \\ C_3 & D_3 \end{bmatrix} = \begin{bmatrix} \cos \Phi_3 & j\frac{Z_s}{Z_0} \sin \Phi_3 \\ j\frac{Z_0}{Z_s} \sin \Phi_3 & \cos \Phi_3 \end{bmatrix} \tag{3.6}$$

Like the outer core, each equal section of the inner core is denoted by

$$\begin{bmatrix} A_4 & B_4 \\ C_4 & D_4 \end{bmatrix} = \begin{bmatrix} \cos \Phi_4 & j\frac{Z_c}{Z_0} \sin \Phi_4 \\ j\frac{Z_0}{Z_c} \sin \Phi_4 & \cos \Phi_4 \end{bmatrix} \tag{3.7}$$

The inner skin:

$$\begin{bmatrix} A_5 & B_5 \\ C_5 & D_5 \end{bmatrix} = \begin{bmatrix} \cos \Phi_5 & j\frac{Z_s}{Z_0} \sin \Phi_5 \\ j\frac{Z_0}{Z_s} \sin \Phi_5 & \cos \Phi_5 \end{bmatrix} \tag{3.8}$$

The entire wiregrid embedded C-sandwich wall configuration is denoted by

$$\begin{bmatrix} A & B \\ C & D \end{bmatrix} = \begin{bmatrix} A_1 & B_1 \\ C_1 & D_1 \end{bmatrix} \begin{bmatrix} A_2 & B_2 \\ C_2 & D_2 \end{bmatrix} \begin{bmatrix} 1 & 0 \\ jB_G & 1 \end{bmatrix} \begin{bmatrix} A_2 & B_2 \\ C_2 & D_2 \end{bmatrix}$$
$$\begin{bmatrix} A_3 & B_3 \\ C_3 & D_3 \end{bmatrix} \begin{bmatrix} A_4 & B_4 \\ C_4 & D_4 \end{bmatrix} \begin{bmatrix} 1 & 0 \\ jB_G & 1 \end{bmatrix} \begin{bmatrix} A_4 & B_4 \\ C_4 & D_4 \end{bmatrix} \begin{bmatrix} A_5 & B_5 \\ C_5 & D_5 \end{bmatrix} \tag{3.9}$$

Therefore, using Eq. (3.9), power transmission coefficient is given by

$$P_{tr} = |T|^2 = \left[\frac{4}{(A + B + C + D)^2} \right] \tag{3.10}$$

And, the power reflection coefficient is given by

$$P_{rf} = |R|_m^2 = \left[\frac{A + B - C - D}{A + B + C + D} \right]^2 \tag{3.11}$$

3.3 Numerical Results and Discussion

The EM performance parameters are computed for C-sandwich radome wall with wiregrid embedded having optimized core thickness. The analysis is done for perpendicular polarization at normal incidence, 45° and 60° incidence angles. This section depicts the comparative study of EM performance of C-sandwich radome wall with wiregrid embedded with that of C-sandwich wall alone. Figure 3.2a shows power transmission characteristics of wiregrid embedded C-sandwich and C-sandwich alone over the frequency band 2–18 GHz at normal incidence. It is evident from the figures that the power transmission characteristic of C-sandwich alone is poor

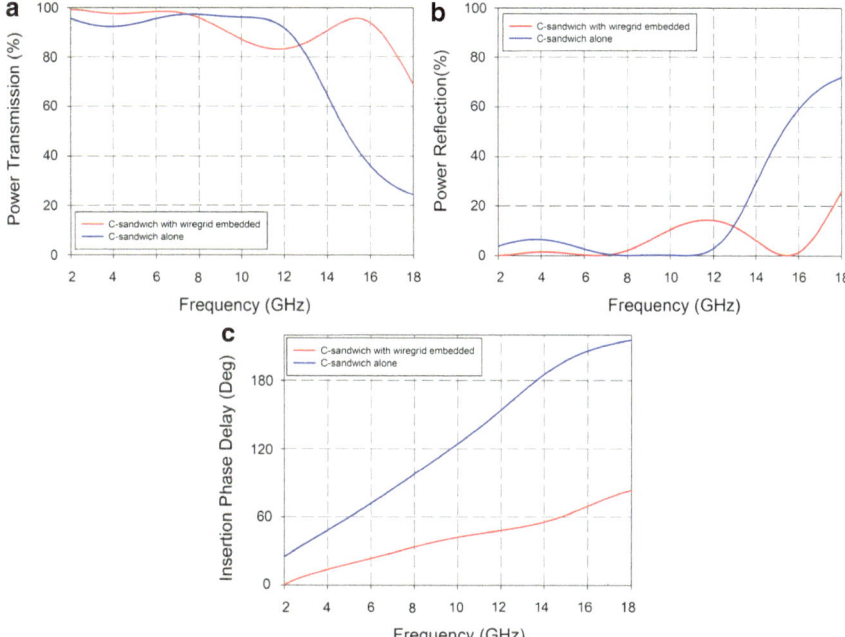

Fig. 3.2 **a** Transmission efficiency of C-sandwich radome wall configurations at normal incidence. **b** Reflection characteristics of C-sandwich radome wall configurations at normal incidence. **c** Insertion phase delay (IPD) of C-sandwich radome wall configurations at normal incidence

as compared to C-sandwich radome wall with wiregrid embedded which is having a considerable transmission efficiency of 95% throughout the frequency band up to 16 GHz, after which it starts degrading and the power transmission ends up to 70% at 18 GHz.

Figure 3.2b shows the power reflection characteristics of the wiregrid embedded C-sandwich and C-sandwich alone at normal incidence. It is observed that the power reflection characteristic of C-sandwich alone at normal incidence is increasing significantly beyond 12 GHz and for C-sandwich with wiregrid embedded the power reflection is reduced considerably.

Figure 3.2c shows the IPD characteristics of modified C-sandwich configuration with wiregrid which characterizes better performance than that of C-sandwich alone at normal incidence.

Figure 3.3a–c shows the power transmission, power reflection, IPD of the wiregrid embedded C-sandwich and C-sandwich alone over the frequency band 2–18 GHz at 45° incidence angle. The characteristics are almost similar, and only slight variations can be noticed as compared to the normal incidence case.

Figure 3.4a–c shows the power transmission, power reflection, IPD of the wiregrid embedded C-sandwich and C-sandwich alone over the frequency range 2–18 GHz at 60° incidence angle.

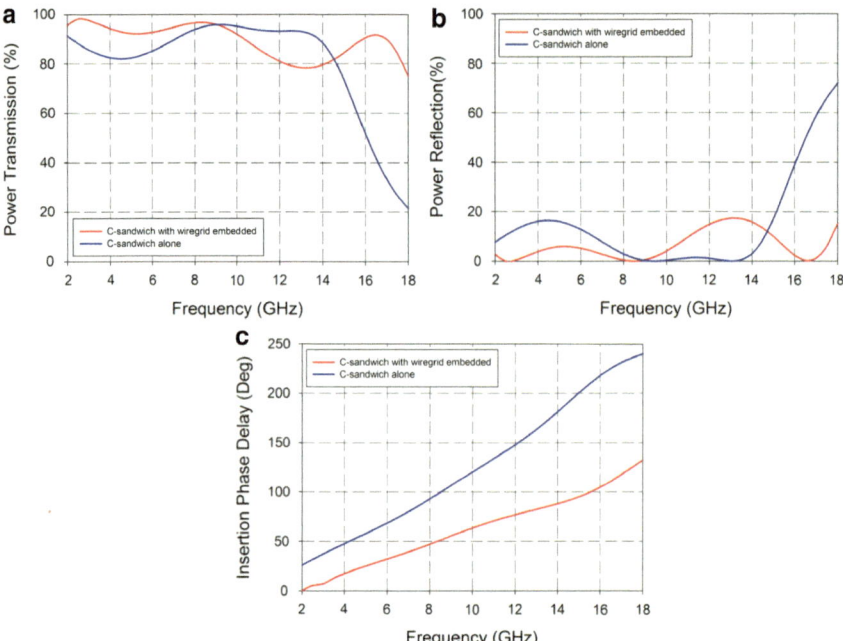

Fig. 3.3 a Transmission efficiency of C-sandwich radome wall configurations at 45° incidence angle. **b** Reflection characteristics of C-sandwich radome wall configurations at 45° incidence angle. **c** Insertion phase delay (IPD) of C-sandwich radome wall configurations at 45° incidence angle

The power reflection characteristics of the C-sandwich radome wall with wiregrid embedded shows excellent performance as compared to the C-sandwich wall alone, which will minimize the side lobe level degradation to a greater extent.

The IPD characteristics of C-sandwich radome wall with wiregrid embedded are better than that of C-sandwich alone at the mentioned three incidence angles. Hence, the modified structure will offer better boresight error (BSE) characteristics.

3.4 Conclusion

The EM performance analysis for wiregrid embedded C-sandwich radome wall with optimized core thickness has been explained in the section. It was found that EM performance characteristics are better for C-sandwich wall configurations with wire-grids embedded. For planar radome applications, C-sandwich wall with wiregrids embedded is a better choice as compared to the conventional C-sandwich radome. The strength-to-weight ratio of the wall configuration can be improved by embedding the wire-grids in the core of the C-sandwich radome.

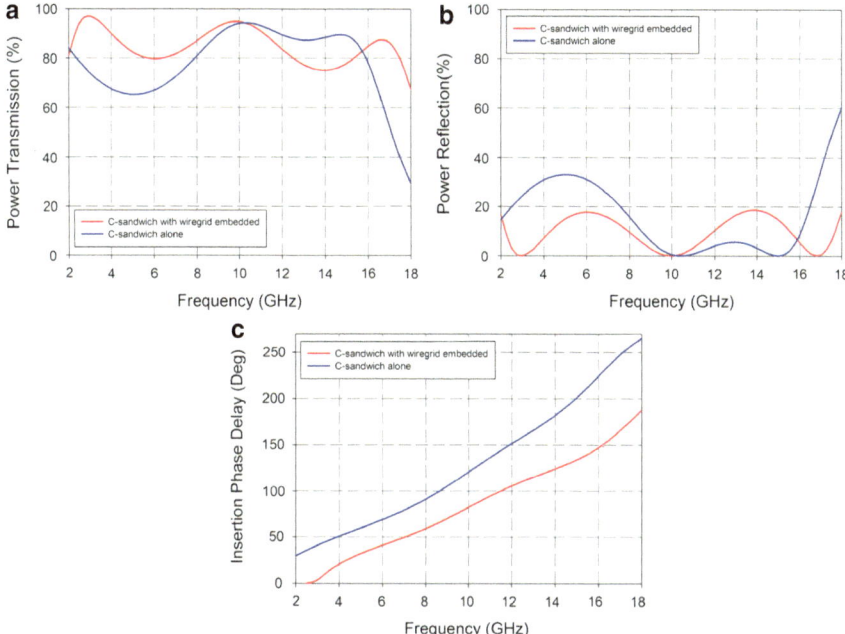

Fig. 3.4 a Transmission efficiency of C-sandwich radome wall configurations at 60° incidence angle. **b** Reflection characteristics of C-sandwich radome wall configurations at 60° incidence angle. **c** Insertion phase delay (IPD) of C-sandwich radome wall configurations at 60° incidence angle

References

Cary RHJ (1983) Radomes. In: Rudge AW, Milne K, Olver AD, Knight P (eds.) The handbook of antenna design. Peter Peregrinus, London, UK. ISBN 0-906048-82-6, 960 p

Harder JF, Gaurino WP (1964) Wide band ECM radomes on high performance aircraft. In: 7th electromagnetic window symposium. Columbus, pp 36–39

Kay AF (1958) Metal inclusions in radomes. Proc OSU-WADC Radome Symp 1:273–290

Miller BM, Brown DL (1957) Grids for Radomes. In: Proceedings of OSU-WADC radome symposium, vol 1, Goodyear Aircraft Corp., pp 138–149, June 1957

Richmond JH (1964) On the design of one layer and three-layer tunable radomes.The Ohio State University, Department of Electrical Engineering, Antenna Laboratory, Report 1751-4 (ASTIA Document No. AD 433839), 15 Jan 1964

Robinson LA (1958) Composite dielectric-metal radomes. Proc OSU-WADC Radome Symp 2:112–131

Tricoles G (1958) Two synthetic approaches to transmission through dielectric sheets. In: Proceedings of OSU-WADC radome symposium, vol 1. Convair, pp 291–323

Walton JD (1966) Techniques for airborne radome design, AFAL report, no. 45433. Wright-Patterson AFB, pp 105–108

Worthington HR (1958) Design of metal screens to improve radome wall performance, NADC-EL-5874 report, Sep 1958

Chapter 4
Broadbanding Techniques Based on Metallic Strip Gratings

Few works have been reported regarding the application of metallic strip gratings for radomes (Primich 1957). In the work reported elsewhere (Schaufelberger 1964), thick metallic gratings were used for the suppression of cross-polarization of radomes. A periodic array of conducting inclusions centrally loaded in the monolithic radome configurations facilitate large bandwidth compared to conventional designs based on identical material parameters and dimensions (Frenkel 2001). Due to high structural rigidity, A-sandwich radomes are generally used in airborne applications as compared to monolithic wall configurations (Kozakoff 2010). To meet the necessities of present-day radome applications, the EM performance of conventional A-sandwich radome may not be adequate. Consequently, in this section, novel designs based on metallic strip gratings are utilized to upgrade the EM performance of A-sandwich radome for airborne applications (Nair et al. 2013). Since metallic strips included in the current work are thin (thickness = 0.1 mm), it is easy to insert them in the layers of A-sandwich wall.

To achieve superior power transmission in the frequency range (S-Band to X-Band), the thickness of core is optimized. Planar array of strips is either fixed on the surfaces or inserted in between the layers for achieving a superior electromagnetic performance. Different configurations used are: (i) metallic strip gratings inserted in between of the mid-plane of the core layer (*Configuration 1*) (Fig. 4.1a) and (ii) metallic strip gratings inserted in each skin–core interface (*Configuration 2*) (Fig. 4.1b).

The ideal design parameters of the metallic strip gratings are found out based on the equivalent transmission line model method for the above-mentioned configurations. The optimum parameters are given in Tables 3.1 and 4.1.

Fig. 4.1 a Cross section of A-sandwich radome panel with metallic strip gratings centrally loaded in the mid-plane of the core (Configuration 1). **b** Cross section of A-sandwich radome panel with metallic strip gratings embedded in each skin core interface (Configuration 2)

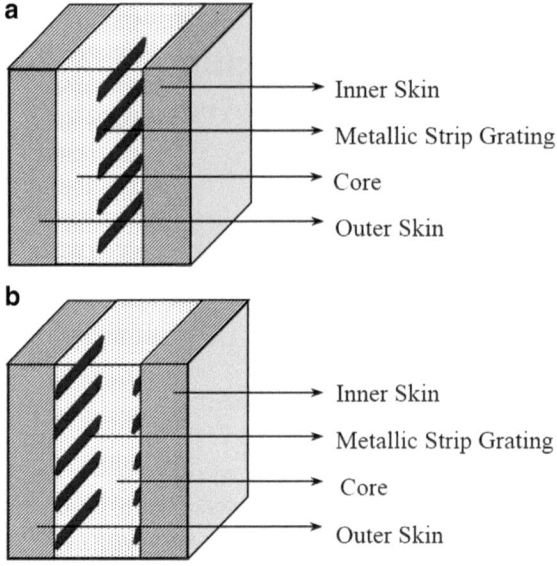

a

Inner Skin

Metallic Strip Grating

Core

Outer Skin

b

Inner Skin

Metallic Strip Grating

Core

Outer Skin

Table 4.1 Design parameters of metallic strip gratings (Configuration 1)

Angle of incidence (°)	Polarization	Optimum width (mm)	Optimum pitch (mm)
0	–	2.6	3.6
45	Perpendicular	2.6	4.0
60	Perpendicular	2.6	4.2
80	Perpendicular	2.6	4.4

4.1 EM Design Aspects of A-sandwich Radome Wall Configuration with Metallic Strip Gratings

For incidence angles from 0° to 80°, power transmission, power reflection, and IPD are evaluated for both configurations mentioned in the section with perpendicular polarization and a comparison is done with A-sandwich alone structure.

The three-layer A-sandwich radome wall configuration is made out of outer skin, core, and inner skin. The material used for outer and inner skin layers is glass–epoxy (dielectric constant, $\varepsilon_r = 4$; and loss tangent, $\tan\delta_e = 0.015$) with constant thickness of 0.75 mm. The center core material used is polyurethane foam ($\varepsilon_r = 1.15$ and $\tan\delta_e = 0.005$) with optimum thickness of 5.44 mm for maximum performance over S-Band to X-Band comprising of frequency range 2–12 GHz. A planar array of thin parallel strips of uniform dimensions is used for the construction of metallic strip gratings with a thickness of 0.1 mm.

The wavelengths corresponding to lower band edge (2 GHz) and higher band edge (12 GHz) are 150 mm and 25 mm, respectively. The optimized core thickness

Table 4.2 Design parameters of metallic strip gratings (Configuration 2)

Angle of incidence (°)	Polarization	Optimum width (mm)	Optimum pitch (mm)
0	–	2.4	2.5
45	Perpendicular	2.4	2.5
60	Perpendicular	2.4	2.5
80	Perpendicular	2.4	2.5

considered here in the section is less than a quarter wavelength corresponding to the above-mentioned lower and upper frequency limits of X-band. Since the thickness of the dielectric layer is less than quarter wavelength, the susceptance will be purely capacitive. The addition of metallic structure acts as a inductive loading which can cancel the effect of capacitive susceptance, thereby enhancing the EM performance for a wider bandwidth.

The EM design strategy involves the optimization of design parameters (pitch and width, keeping thickness constant) of the metallic strip grating in such a way that that the inductive susceptance of strip grating matches with the capacitive susceptance of the entire A-sandwich radome wall. Hence, the resulting structure acts as spatial low-pass filter, providing superior EM performance over the entire frequency band (Table 4.2).

The metallic strip gratings are very sensitive to polarization and angle of incidence. For parallel polarization, the EM performance characteristics are better as compared to the perpendicular polarization for a radome with identical dimensions and same material. It happens because the wave impedance for parallel polarization is less than that of perpendicular polarization for a particular incidence angle which accounts for the performance degradation in the perpendicular polarization. Considering these, the perpendicular polarization is investigated in the present work.

4.2 EM Performance Analysis of A-sandwich Radome Wall Configuration with Metallic Strip Gratings

An equivalent transmission line model can be derived from the metallic strip grating included A-sandwich radome. The variation in the characteristic impedance from free space to A-sandwich wall is a major source for incident wave power reflection on the structure. An end-to-end low impedance line can be derived from different layers of radome wall as compared to the free space. Hence, an equivalent matrix can be derived from the cascaded individual layers by multiplication of individual layer matrices.

For configuration 1 (Fig. 4.1a), the matrix representation of each layer of A-sandwich wall is given by:

The inner skin is represented by

$$\begin{bmatrix} A_1 & B_1 \\ C_1 & D_1 \end{bmatrix} = \begin{bmatrix} \cos \Phi_1 & j\frac{z_s}{z_0} \sin \Phi_1 \\ j\frac{z_0}{z_s} \sin \Phi_1 & \cos \Phi_1 \end{bmatrix} \tag{4.1}$$

In Configuration 1, the metallic strip grating is located at the center plane of the core. Hence, the strip grating is embedded in the center plane of the core; it can be divided into two similar sections. Matrix form of one half-section is given by

$$\begin{bmatrix} A_2 & B_2 \\ C_2 & D_2 \end{bmatrix} = \begin{bmatrix} \cos \Phi_2 & j\frac{z_c}{z_0} \sin \Phi_2 \\ j\frac{z_0}{z_c} \sin \Phi_2 & \cos \Phi_2 \end{bmatrix} \tag{4.2}$$

Let A_{MS}, B_{MS}, C_{MS}, and D_{MS} be the elements of the matrix belonging to metallic strip grating

$$\begin{bmatrix} A_{MS} & B_{MS} \\ C_{MS} & D_{MS} \end{bmatrix} = \begin{bmatrix} 1 & 0 \\ jB_G & 1 \end{bmatrix} \tag{4.3}$$

Here, B_G represents the shunt susceptance of the metallic strip grating (Lee et al.1985). For perpendicular polarization, the parallel susceptance of the strip for perpendicular polarization is given by

$$B_G = \frac{-4p \sec \theta}{\lambda} \left[\left(\ln \ \mathrm{cosec} \ \left(\frac{\pi g}{2p} \right) + G \right) \right] \tag{4.4}$$

Here, the correction term is given by

$$G = \frac{0.5\left(1 - \beta^2\right)^2 \left[\left(1 - \frac{\beta^2}{4}\right)(C_{1+} + C_{1-}) + 4\beta^2 C_{1+} C_{1-} \right]}{\left(1 - \frac{\beta^2}{4}\right) + \beta^2 \left(1 + \frac{\beta^2}{2} - \frac{\beta^4}{8}\right)(C_{1+} + C_{1-}) + 2\beta^6 C_{1+} C_{1-}} \tag{4.5}$$

The coefficients are given by

$$C_{1\pm} = \frac{1}{\sqrt{\left(\frac{p \sin \theta}{\lambda} \pm 1\right)^2 - \frac{p^2}{\lambda^2}}} - 1 \tag{4.6}$$

Here, $\beta = \sin\left(\frac{\pi w}{2p}\right)$ and $n = \pm 1, \pm 2, \pm 3 \ldots$
The other half-section of the core is represented by

$$\begin{bmatrix} A_3 & B_3 \\ C_3 & D_3 \end{bmatrix} = \begin{bmatrix} \cos \Phi_3 & j\frac{z_c}{z_0} \sin \Phi_3 \\ j\frac{z_0}{z_c} \sin \Phi_3 & \cos \Phi_3 \end{bmatrix} \tag{4.7}$$

The outer skin is represented by

$$\begin{bmatrix} A_4 & B_4 \\ C_4 & D_4 \end{bmatrix} = \begin{bmatrix} \cos \Phi_4 & j\frac{z_s}{z_0} \sin \Phi_4 \\ j\frac{z_0}{z_s} \sin \Phi_4 & \cos \Phi_4 \end{bmatrix} \tag{4.8}$$

Thus, the equivalent matrix representation for the A-sandwich radome with Configuration 1 is given as

$$\begin{bmatrix} A & B \\ C & D \end{bmatrix} = \begin{bmatrix} A_1 & B_1 \\ C_1 & D_1 \end{bmatrix}\begin{bmatrix} A_2 & B_2 \\ C_2 & D_2 \end{bmatrix}\begin{bmatrix} 1 & 0 \\ jB_G & 1 \end{bmatrix}\begin{bmatrix} A_3 & B_3 \\ C_3 & D_3 \end{bmatrix}\begin{bmatrix} A_4 & B_4 \\ C_4 & D_4 \end{bmatrix} \tag{4.9}$$

And the equivalent matrix representation for the A-sandwich radome for Configuration 2 with metallic strip grating incorporated in each skin–core interface (Fig. 4.1b) is given as

$$\begin{bmatrix} A & B \\ C & D \end{bmatrix} = \begin{bmatrix} A_1 & B_1 \\ C_1 & D_1 \end{bmatrix}\begin{bmatrix} 1 & 0 \\ jB_G & 1 \end{bmatrix}\begin{bmatrix} A_2 & B_2 \\ C_2 & D_2 \end{bmatrix}\begin{bmatrix} 1 & 0 \\ jB_G & 1 \end{bmatrix}\begin{bmatrix} A_4 & B_4 \\ C_4 & D_4 \end{bmatrix} \tag{4.10}$$

Here, $\begin{bmatrix} A_2 & B_2 \\ C_2 & D_2 \end{bmatrix}$ represents the core of the radome wall, which is different from that of Configuration 1.

The A, B, C, and D parameters of the final matrix of each configuration are calculated using Eqs. (4.9) and (4.10). The power transmission and power reflection coefficients are evaluated by Eqs. (3.10) and (3.11).

The IPD of the radome wall is investigated for determining the phase distortions. The IPD for Configuration 1 is given by

$$\text{IPD}_1 = -\angle T_1 - \frac{2\pi}{\lambda}(2t_s + 2t_{c1} + t_{ms})\cos\theta \tag{4.11}$$

Similarly, IPD for Configuration 2 is given by

$$\text{IPD}_2 = -\angle T_2 - \frac{2\pi}{\lambda}(2t_s + t_{c2} + 2t_{ms})\cos\theta \tag{4.12}$$

Here, $\angle T_1$ and $\angle T_2$ represent the phase angles for Configuration 1 and Configuration 2, respectively. The thicknesses of skin layers and metallic strip grating are given by t_s and t_{ms}, respectively. Let t_{c1} be the thickness of each section of the core for Configuration 1, while t_{c2} be the core thickness for Configuration 2.

4.3   Numerical Results and Discussion

A relative investigation of the EM performance of A-sandwich wall with metallic strip gratings for Configurations 1 and 2 and A-sandwich wall alone is done. For the entire frequency band of 2–12 GHz, the EM performance parameters are calculated at normal incidence, 45°, 60°, and 80° for perpendicular polarization. The EM performance characteristics of Configuration 1 are shown in Figs. 4.2a–c, 4.3a–c, 4.4a–c, 4.5a–c. It is evident that the metallic strip grating included A-sandwich wall gives better performance over A-sandwich wall alone case (Figs. 4.2a, 4.3a, 4.4a, 4.5a). A-sandwich wall embedded with the strip grating gives a superior power transmission which is well above 90% at normal incidence, 45°, and 60° incidence angles. However, the power transmission efficiency degrades for high incidence angles (80°). The power reflection for A-sandwich radome wall with both configurations (1 and 2) is less as compared to the A-sandwich wall alone (Figs. 4.2b, 4.3b, 4.4b, 4.5b). The power reflection of the conventional A-sandwich structure is more for higher incidence angle. The IPD characteristics are shown in Figs. 4.2c, 4.3c, 4.4c, 4.5c. At normal incidence, the IPD of the strip embedded A-sandwich wall is similar to the conventional radome structure. However, as the incidence angle increases, the IPD of the strip embedded structure also will increase.

The power transmission, power reflection, and IPD characteristics of an A-sandwich radome wall with Configuration 2 mentioned in the section and A-sandwich wall alone are plotted and shown in Figs. 4.6a–c, 4.7a–c, 4.8a–c, 4.9a–c. The optimized design parameters of metallic strip gratings are given in Table 4.1. The power transmission characteristics of Configuration 2 at 0°, 45°, 60°, and 80° incidence angles are plotted in Figs. 4.6a, 4.7a, 4.8a, 4.9a.The presence of metallic strip gratings at each skin–core interface enhances the power transmission efficiency of the Radomes (above 90% at normal incidence). For other incidence angles, (>0°) a decrease in power transmission efficiency is observed. Nevertheless across the frequency band of 2–12 GHz, the power transmission characteristics of A-sandwich wall with strip gratings are superior as compared to the A-sandwich alone at various incidence angles. The power reflection characteristics of Configuration 2 are plotted in Figs. 4.6b, 4.7b, 4.8b, 4.9b. The A-sandwich wall embedded with metallic strip gratings shows very low power reflection as compared to the conventional structure. The power reflection for Configuration 2 is very low for normal incidence, 45°, and 60° incidence angles. However, in conventional structures, the power reflection significantly increases at a high incidence angle of 80°. The IPD characteristics of Configuration 2 are shown in Fig. (4.6c, 4.7c, 4.8c, 4.9c). It is evident from the figures that the IPD of the A-sandwich wall with Configuration 2 is superior than that of A-sandwich alone, thereby lowering the boresight error and phase distortions.

The Configuration 1 has superior EM performance characteristics as compared to Configuration 2 with power transmission efficiency of 85% and 82%, respectively. Also, the average power reflection for Configuration 1 is nearly zero even at high incidence angle 80°, whereas it is around 5% for Configuration 2. Low power reflection characteristic of Configuration 1 reduces antenna side lobe level (SLL) degradations and generation of flash lobes. IPD of Configuration 1 is greater than that of Configuration 2 for all incidence angles, which indicates more phase distortions and boresight error.

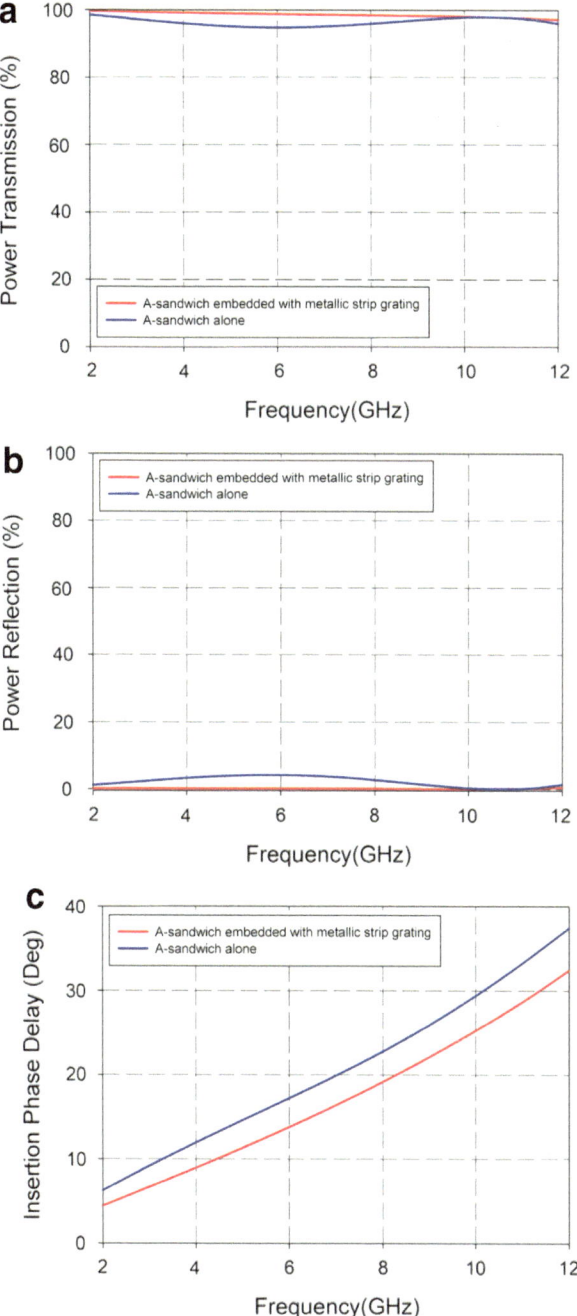

Fig. 4.2 a Transmission efficiency of A-sandwich radome wall configurations at normal incidence. **b** Reflection characteristics of A-sandwich radome wall configurations at normal incidence.
c Insertion phase delay (IPD) of A-sandwich radome wall configurations at normal incidence

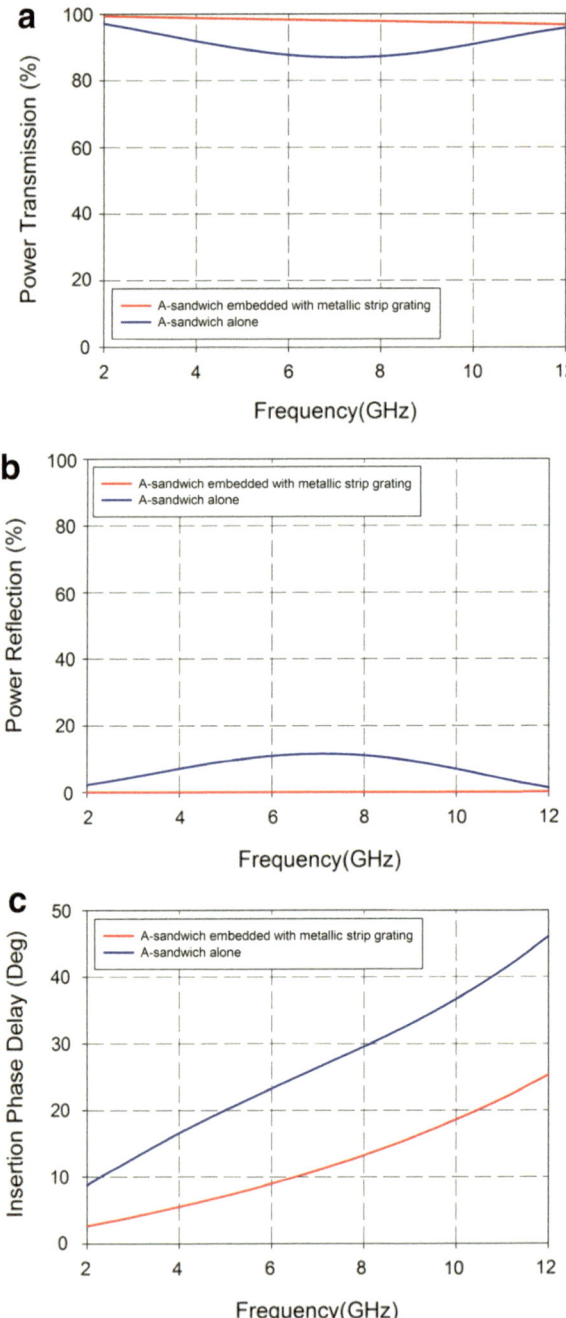

Fig. 4.3 **a** Transmission efficiency of A-sandwich radome wall configurations at 45° incidence angle. **b** Reflection characteristics of A-sandwich radome wall configurations at 45°. **c** Insertion phase delay of A-sandwich radome wall configurations at 45°

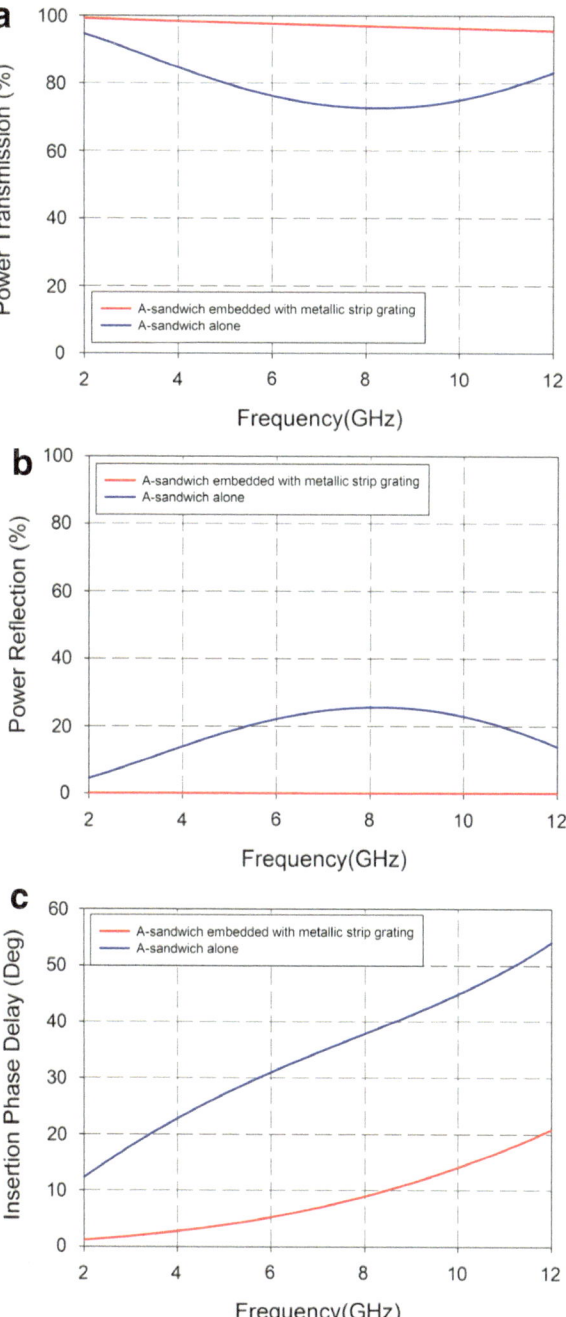

Fig. 4.4 a Transmission efficiency of A-sandwich radome wall configurations at 60°. **b** Reflection characteristics of A-sandwich radome wall configurations at 60°. **c** Insertion phase delay of A-sandwich radome wall configurations at 60°

Fig. 4.5 **a** Transmission
efficiency of A-sandwich
radome at 80°. **b** Reflection
characteristics of
A-sandwich radome wall
configurations at 80°.
c Insertion phase delay of
A-sandwich radome wall
configurations at 80°
incidence angle

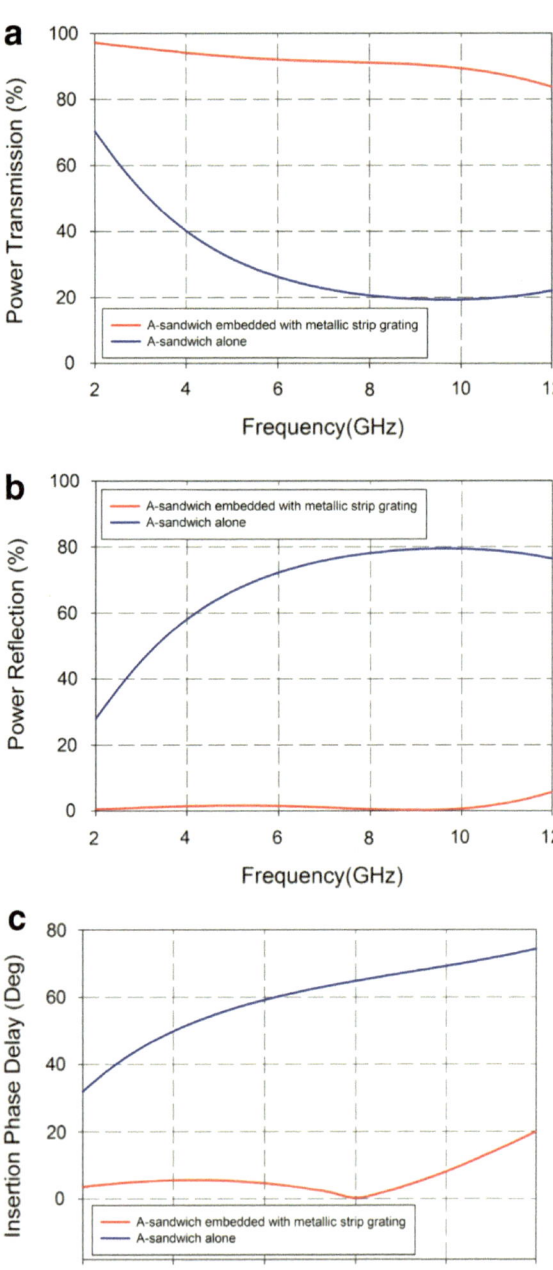

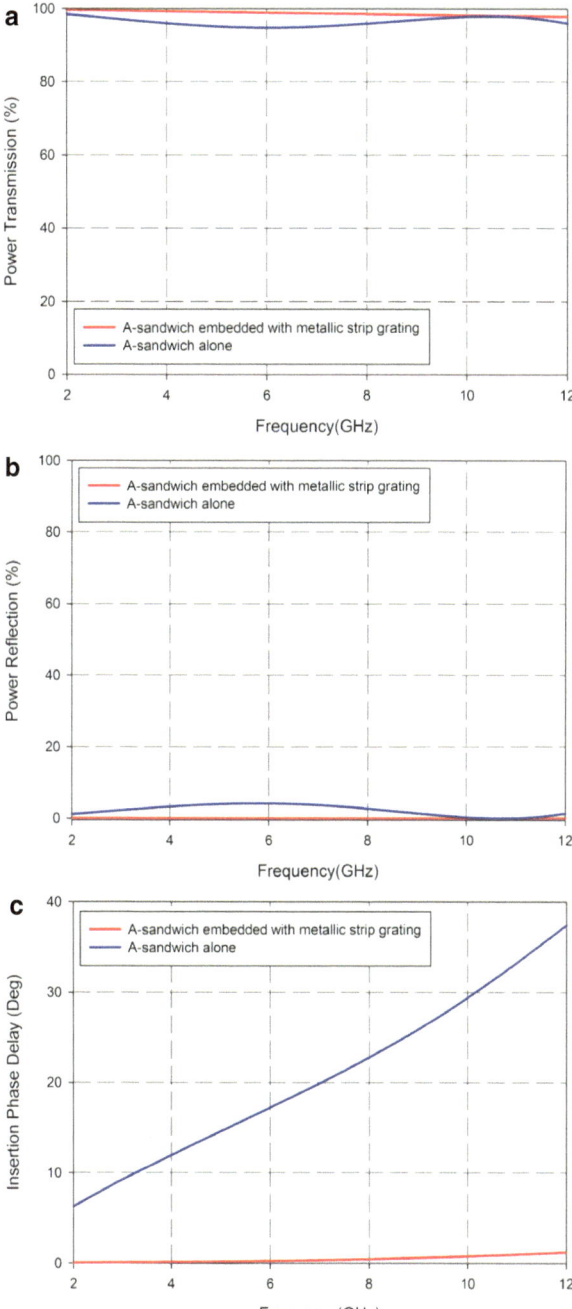

Fig. 4.6 **a** Transmission efficiency of A-sandwich radome wall configurations at normal incidence. **b** Reflection characteristics of A-sandwich radome wall configurations at normal incidence.
c Insertion phase delay of A-sandwich radome wall configurations at normal incidence

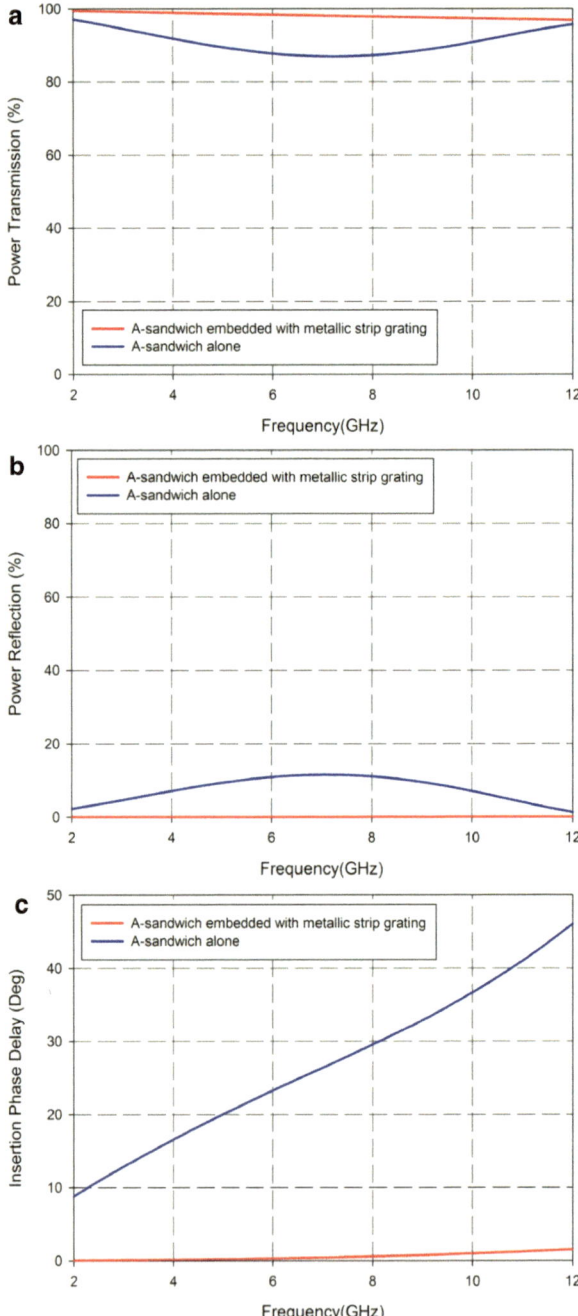

Fig. 4.7 **a** Transmission efficiency of A-sandwich radome wall configurations 45°. **b** Reflection characteristics of A-sandwich radome wall configurations at 45°. **c** Insertion phase delay of A-sandwich radome wall configurations at 45°

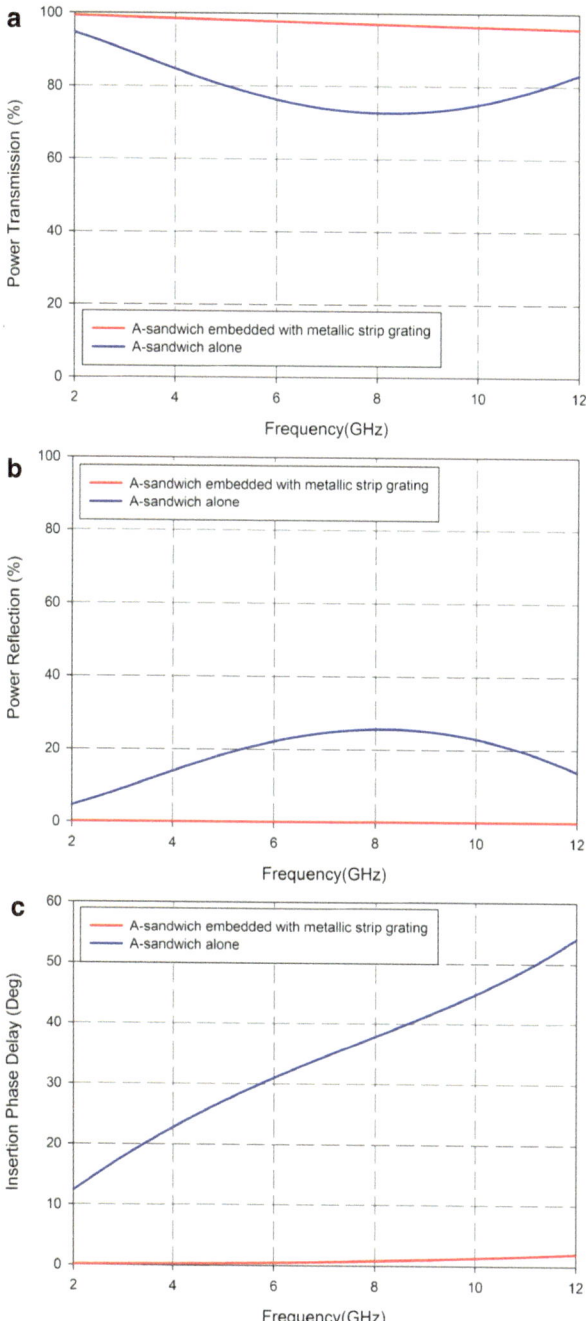

Fig. 4.8 **a** Transmission efficiency of A-sandwich radome wall configurations at 60°. **b** Reflection characteristics of A-sandwich radome wall configurations at 60°. **c** Insertion phase delay of A-sandwich radome wall configurations at 60°

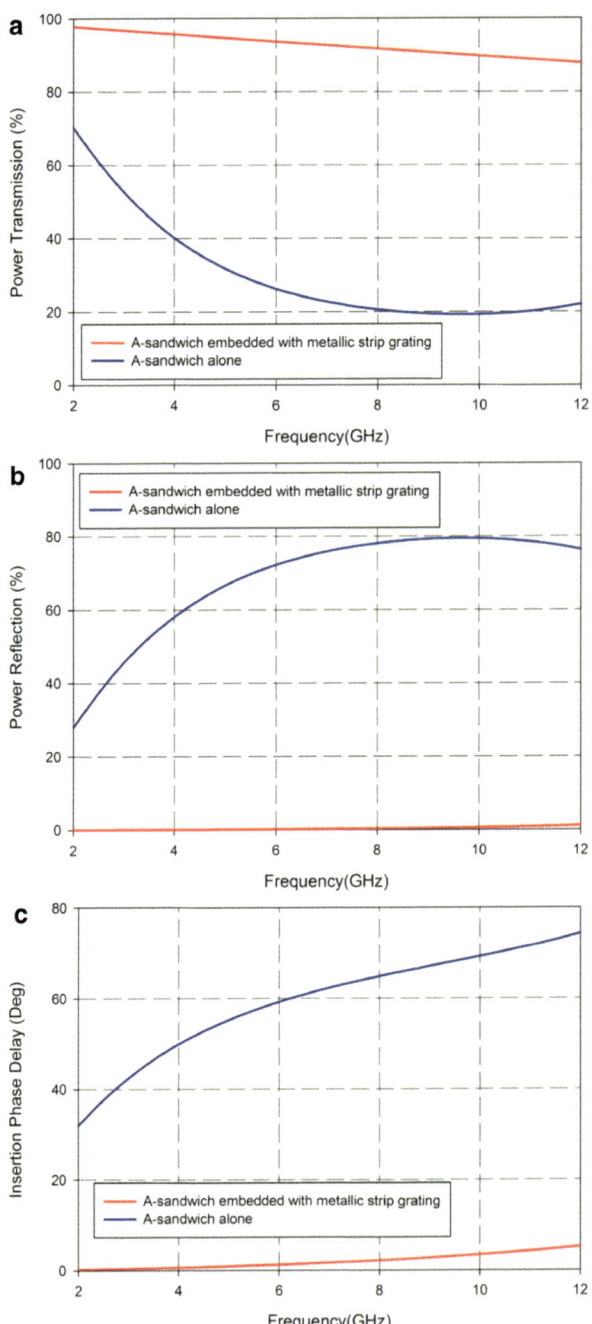

Fig. 4.9 **a** Transmission efficiency of A-sandwich radome wall configurations at 80°. **b** Reflection characteristics of A-sandwich radome wall configurations at 80°. **c** Insertion phase delay of A-sandwich radome wall configurations at 80°

4.4 Conclusion

This chapter showcases the importance of metallic strip gratings in enhancing the EM performance characteristics of A-sandwich radome across a frequency band of 2–12 GHz. It was found that EM performance characteristics are better for A-sandwich wall configurations with embedded metallic strip gratings (Configuration 1 and 2) over conventional A-sandwich alone wall. Among both configurations, *Configuration* 1 is preferred to *Configuration* 2 in terms of EM performance. For planar radome applications, A-sandwich wall with metallic strip gratings is a better choice as compared to the conventional A-sandwich radome. From fabrication point of view, *Configuration* 1 is preferable as only one set of strip grating has to be embedded in the structure.

Reference

Frenkel A (2001) Thick metal dielectric radome. IET Electron Lett 38:1374–1375

Kozakoff DJ (2010) Analysis of radome enclosed antennas. Artech House, Norwood. ISBN 1596934417, 294 p

Nair RU, Madhumitha J, Jha RM (2013) Application of metallic strip gratings for enhancement of electromagnetic performance of A-sandwich radome. Defense Sci J 63:508–514

Primich RJ (1957) Some reflection and transmission properties of a strip grating. IRE Trans Antennas Propag AP-5:176–182

Schaufelberger AH (1964) Antenna radome gratings for conical scan systems. In: Proceedings of OSU-RTD symposium on electromagnetic windows, vol 2, session-III, pp 1–18, June 1964

Chapter 5
Broadbanding Techniques Based on Frequency Selective Surfaces (FSS)

Radome wall configurations have to be modified in order to attain better performance parameters such as high transmission efficiency, low cross-polarization levels, low boresight error, etc., which is an arduous task. Conventional methods followed to obtain better EM performance parameters over a broadband of frequencies are use of thin parallel metallic wire arrays and wire meshes as detailed before in previous chapters. Conventional monolithic radome panels when loaded periodically with metallic inclusions offered much wider bandwidth as discussed in previous chapters.

While dielectric walls embedded with perforated metallic sheets or sheets fixed on back side of radome wall in some applications offered narrow bandwidth and also they give better results at high incidence angles.

Enhancing the transmission efficiency of the radome walls engrossed the main focus of most of the works reported earlier (Cary 1983; Kozakoff 2010; Walton 1966). Radome walls with embedded metallic structures improve some of the performance parameters at the cost of degradation of some other parameters. Hence, this demands novel techniques such as FSS radomes which provide an enhancement in all EM performance parameters simultaneously. (Munk 2005; Vardaxoglou 1997; Wu 1992). In most of the reported works, FSS layers were embedded in the radome wall structures to achieve broadband transmission characteristics.

5.1 Broadbanding Techniques Based on Single Square Loop Frequency Selective Surfaces (SSL-FSS)

Transcendent EM performance parameters at the frequency of operation as well as the structural rigidity for enduring aerodynamic conditions at high mach speeds make monolithic half-wave radome walls greatly preferable for streamlined nosecone radomes. Even then, new designs are indispensable due to the stringent requirements such as multiple bands and higher bandwidths, which monolithic could not deliver.

© The Author(s), under exclusive license to Springer Nature Singapore Pte Ltd. 2020   33
P. S. Mohammed Yazeen et al., *Broadbanding Techniques for Radomes*,
SpringerBriefs in Computational Electromagnetics,
https://doi.org/10.1007/978-981-33-4130-2_5

Fig. 5.1 a Schematic of the single square loop (SSL)-FSS. Here *p*—periodicity; d_{FSS}—thickness of SSL-FSS; *g*—gap length; *d*—length of SSL element; and *w*—width of SSL element. **b** Monolithic slab of half-wave thickness. SSL-FSS is centrally loaded *w.r.t.* the mid-plane

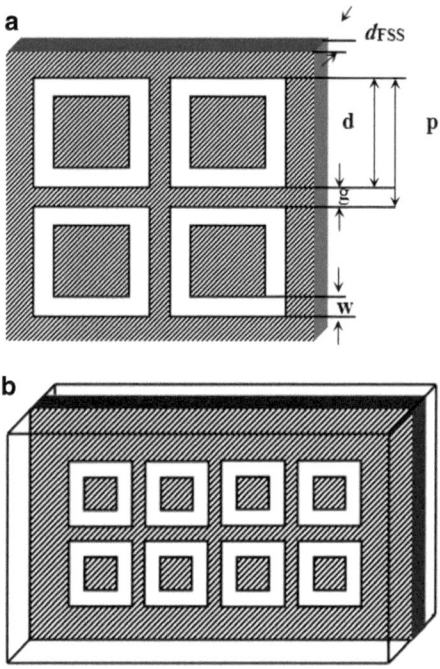

A monolithic wall configuration embedded with SSL-FSS is able to deliver superior EM performance parameters in 1–12 GHz frequency band (L to X band), compared to simple monolithic half-wave wall at various incidence angles.

An aperture-type SSL-FSS for broadbanding of monolithic half-wave radome wall configuration is presented in Fig. 5.1a, b.

5.1.1   EM Design Aspects of SSL-FSS Embedded Monolithic Radome.

Glass epoxy material with permittivity 4.0, loss tangent 0.015, and wall thickness (d_m) of 7.44 mm is chosen for the monolithic half-wave wall radomes at 10 GHz for better power transmission at normal incidence. Since the half-wave radome wall thickness is much higher at lower frequencies, SSL-FSS is opted for reducing the payload. SSL-FSS is where the half-wave radome walls are centrally loaded with periodic single square loop aperture arrays, as illustrated in Fig. 5.2a.

The whole wall configuration of SSL-FSS embedded monolithic radome wall can be represented by an equivalent transmission line as shown in Fig. 5.2b, where different sections are corresponding to slab and SSL-FSS. Different layers are considered as low impedance lines cascaded, and the whole configuration is the product

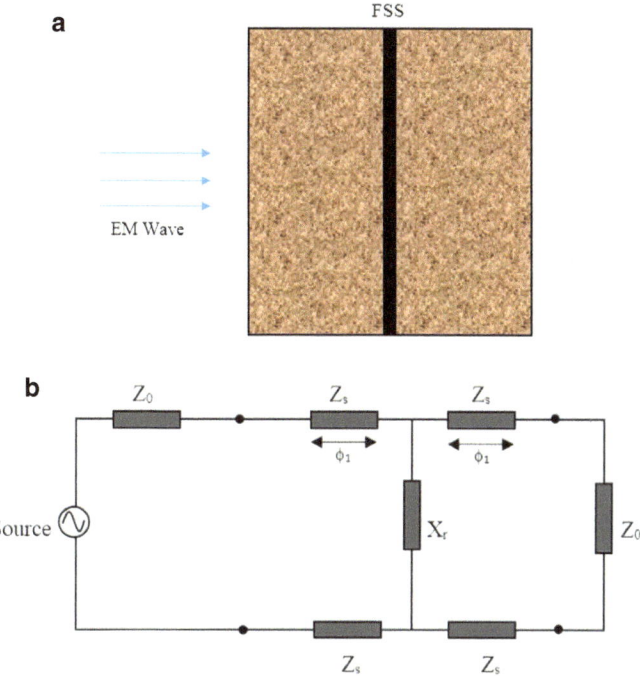

Fig. 5.2 **a** Lateral view of monolithic half-wave slab centrally loaded with SSL-FSS. **b** Equivalent circuit of SSL-FSS embedded radome wall

of all corresponding individual matrices. Variation in characteristic impedance from free space to radome wall interface causes reflection of power on the structure.

Design parameters of the SSL-FSS structure for the entire frequency band from S to X band are optimized and included in Table 5.1 at various angles of incidence. In this design, the width w and d_{FSS} are not varied due to practical considerations. Parameters d, g, and p are optimized to obtain better radome performance parameters.

Table 5.1 Design parameters of SSL-FSS

Angle of incidence θ (in degrees)	Length, d (mm)	Width, w (mm)	Gap length, g (mm)	Period, p (mm)	Thickness, d_{FSS} (mm)
0°	13.20	3.0	2.0	15.20	3.0
30°	10.25	3.0	1.0	11.25	3.0
45°	12.00	3.0	1.0	13.00	3.0
60°	17.00	3.0	1.0	18.00	3.0

5.1.2 EM Performance Analysis of SSL-FSS Embedded Monolithic Radome

SSL-FSS is embedded in the mid-plane of monolithic half-wave radome wall, and thus it is considered to be comprising of two identical sections with SSL-FSS in middle of them.

The characteristic impedance of free space is defined as Z_0 and that of each half-section of monolithic half-wave radome wall is given as Z_s. Each layer is represented by corresponding matrices with ABCD parameters in equivalent transmission line method, whereas A_i, B_i, C_i, and D_i represent the ABCD parameters of ith layer. Here, 't' represents the thickness of the half-section of monolithic half-wave wall and d_{FSS} represents the thickness of SSL-FSS, respectively. 't' is given by half the difference between wall thickness (d_m) and SSL-FSS thickness (d_{FSS}).

ABCD matrix of corresponding half-section of the monolithic radome wall is given by

$$\begin{bmatrix} A_1 & B_1 \\ C_1 & D_1 \end{bmatrix} = \begin{bmatrix} \cos \Phi_1 & j\frac{Z_s}{Z_0} \sin \Phi_1 \\ j\frac{Z_0}{Z_s} \sin \Phi_1 & \cos \Phi_1 \end{bmatrix} \tag{5.1}$$

Here, Φ represents the electrical length of the respective layer.
Second half-section of the monolithic radome wall is given by

$$\begin{bmatrix} A_2 & B_2 \\ C_2 & D_2 \end{bmatrix} = \begin{bmatrix} \cos \Phi_2 & j\frac{Z_s}{Z_0} \sin \Phi_2 \\ j\frac{Z_0}{Z_s} \sin \Phi_2 & \cos \Phi_2 \end{bmatrix} \tag{5.2}$$

In the equivalent transmission line method, aperture-type FSS is represented as a transmission line with shunt reactance. The EM characteristics of infinite periodic array are assumed to be same as that of single unit cell of SSL-FSS.

Parameters of the SSL-FSS are optimized to obtain the matching between the inductive reactance offered by the SSL-FSS and capacitive reactance offered by the monolithic radome wall alone to yield better EM performance over broadband of frequencies.

Aperture-type SSL-FSS can be represented as an ABCD matrix, where the elements of the matrix are represented by A_{FSS}, B_{FSS}, C_{FSS}, and D_{FSS}, and it is expressed as shown below.

$$\begin{bmatrix} A_{FSS} & B_{FSS} \\ C_{FSS} & D_{FSS} \end{bmatrix} = \begin{bmatrix} 1 & 0 \\ 1/jX_r & 1 \end{bmatrix} \tag{5.3}$$

In the above equation, X_r represents the reactance of SSL-FSS, whereas inductive reactance X and capacitive susceptance B contribute to X_r. (Ohira et al. 2005).
For TE incidence, X and B are given by

$$X_{\text{TE}} = \frac{p\cos\theta}{\lambda}\left[\ln\left(\frac{1}{\sin(\pi w/2p)}\right)\right] \tag{5.4}$$

$$B_{\text{TE}} = \frac{4p\sec\theta}{\lambda}\left[\ln\left(\frac{1}{\sin(\pi g/2p)}\right)\right] \tag{5.5}$$

In the above expressions, g, p, and w represent the SSL-FSS parameters where g represents the gap width, p represents the periodicity, w represents the width, and λ represents the wavelength of the EM wave being incident.

SSL-FSS is considered to be embedded in the monolithic radome wall to reduce the losses due to mismatch in the permittivity of the medium.

So, the entire monolithic half-wave wall with SSL-FSS is represented by

$$\begin{bmatrix} A & B \\ C & D \end{bmatrix} = \begin{bmatrix} A_1 & B_1 \\ C_1 & D_1 \end{bmatrix}\begin{bmatrix} 1 & 0 \\ \frac{1}{jX_r} & 1 \end{bmatrix}\begin{bmatrix} A_2 & B_2 \\ C_2 & D_2 \end{bmatrix} \tag{5.6}$$

Equation (5.6) is used to compute the final ABCD matrix parameters. The power transmission and power reflection coefficients are given by Eqs. (3.10) and (3.11).

IPD is used to determine the phase distortions of the radome wall. IPD of the SSL-FSS radome wall is given by

$$\text{IPD} = -\angle T - \frac{2\pi}{\lambda}(2t + d_{\text{FSS}})\cos\theta \tag{5.7}$$

In the above equation, $\angle T$ represents the phase angle linked to the transmission coefficient T. In the design of SSL-FSS, parameters d and g are the critical ones, and any trivial change in these parameters causes significant variations in the EM performance parameters.

5.1.3 Numerical Results and Discussion

EM performance parameters at various angles of incidence (0°, 30°, 45°, and 60°, respectively) of monolithic radome wall with SSL-FSS structure are computed. EM performance at incidence angles ranging from 0° to 30° characterizes hemispherical and cylindrically shaped radomes which are normal incidence radomes, 0°–45° range is for paraboloidal radomes, and 0°–60° for the highly streamlined radomes (conical radome, ogival radome, etc.).

At normal incidence, the EM performance characteristics of conventional monolithic radome wall structure and SSL-FSS monolithic radome wall structures are shown in Fig. 5.3a–c. Figure 5.3a compares the power transmission characteristics of SSL-FSS embedded monolithic radome wall and conventional monolithic radome wall over a range of frequencies from 0.5 to 12 GHz at normal angle of incidence.

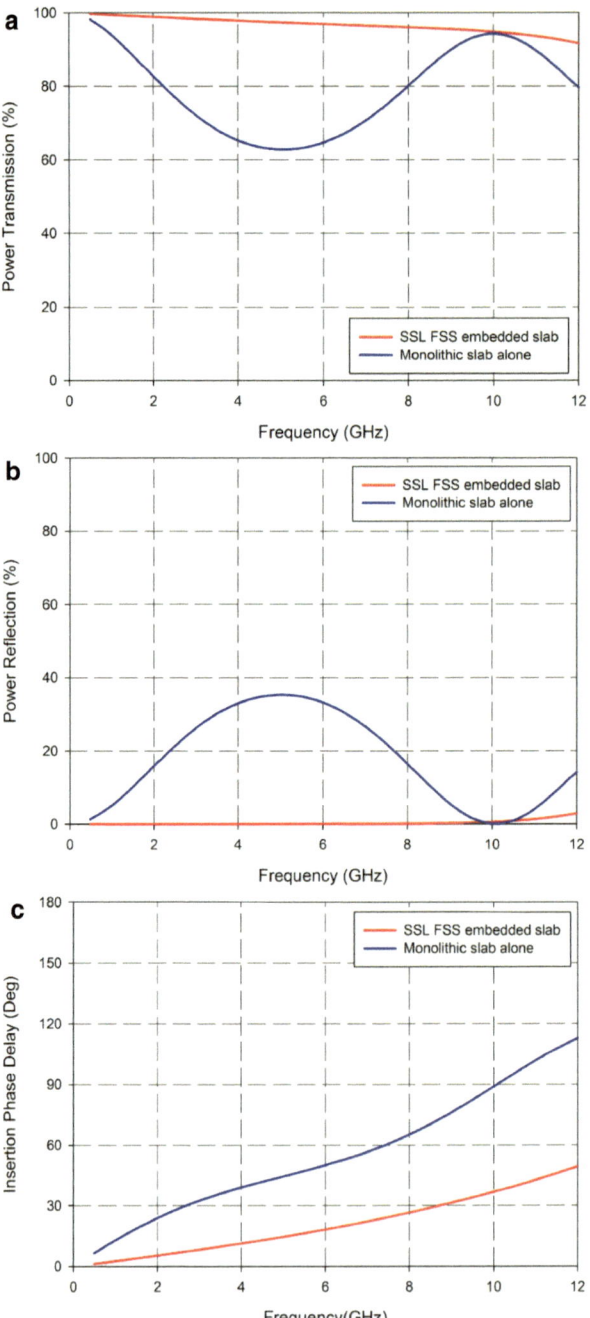

◄Fig. 5.3 **a** Transmission efficiency of SSL-FSS embedded radome and monolithic radome at normal incidence over a broadband (0.5–12 GHz). **b** Reflection characteristics of SSL-FSS embedded radome and monolithic radome at normal incidence over a broadband (0.5–12 GHz). **c** Insertion phase delay of SSL-FSS embedded radome and monolithic radome at normal incidence over a broadband (0.5–12 GHz)

Power transmission efficiency of conventional monolithic radome wall alone is around 95% at 0.5 GHz for normal angle of incidence, then decreases gradually and drops to a minimum around 5 GHz, and again increases to maximum around 10 GHz, which is the design frequency of the monolithic radome wall.

Thereafter, the power transmission characteristics of SSL-FSS embedded monolithic radome wall structure are studied, and it shows better performance parameters over a wide frequency band as compared to monolithic wall alone structure. Even then, the transmission efficiency declines gradually from 0.5 to 12 GHz in the SSL-FSS embedded radome wall case.

In the case of SSL-FSS embedded monolithic radome wall, the power reflection is considerably low in the frequency range from 0.5 to 10 GHz, it is shown in Fig. 5.3b, and it is a significant performance. As in the case of power transmission characteristics, here power reflection increases from 0.5 GHz, reaches to a maximum of 35% at 5 GHz, then decreases to a minimum value at 10 GHz, and then again increases to maximum after that.

Figure 5.4a–c depicts the EM performance characteristics of SSL-FSS embedded dielectric structures and monolithic radome wall alone structures at an incidence angle of 30°. Here also, the power transmission efficiency of SSL-FSS embedded radome wall is far better than that of monolithic radome wall alone structure. However, due to the change in angle of incidence, power transmission efficiency in this case is also reduced for SSL-FSS embedded radome.

Similar to normal incidence case, SSL-FSS embedded radome panels exhibit almost negligible power reflection characteristics at 30° incidence angle over a broadband of frequencies, which makes this structure appropriate for radome applications. Monolithic radome wall alone structure exhibits an oscillatory nature for power reflection. Reduced level of reflected power helps in side lobe level suppression and elimination of flash lobes.

Similarly, IPD of SSL-FSS embedded radome panel holds better values than that of monolithic radome wall structure alone.

Figure 5.5a–c illustrates the EM performance parameters of SSL-FSS embedded radome panel and monolithic radome wall alone structures at an incidence angle of 45°. In this case as well power transmission, power reflection and IPD exhibited by SSL-FSS are much better compared to monolithic radome wall alone structure in the broadband of frequencies.

Similarly, Fig. 5.6a–c illustrates the EM performance parameters of SSL-FSS embedded dielectric radome and monolithic radome wall alone structures at an incidence angle of 60°. As the incidence angle is increased to a much higher value, the power transmission efficiency of SSL-FSS embedded radome panel is reduced

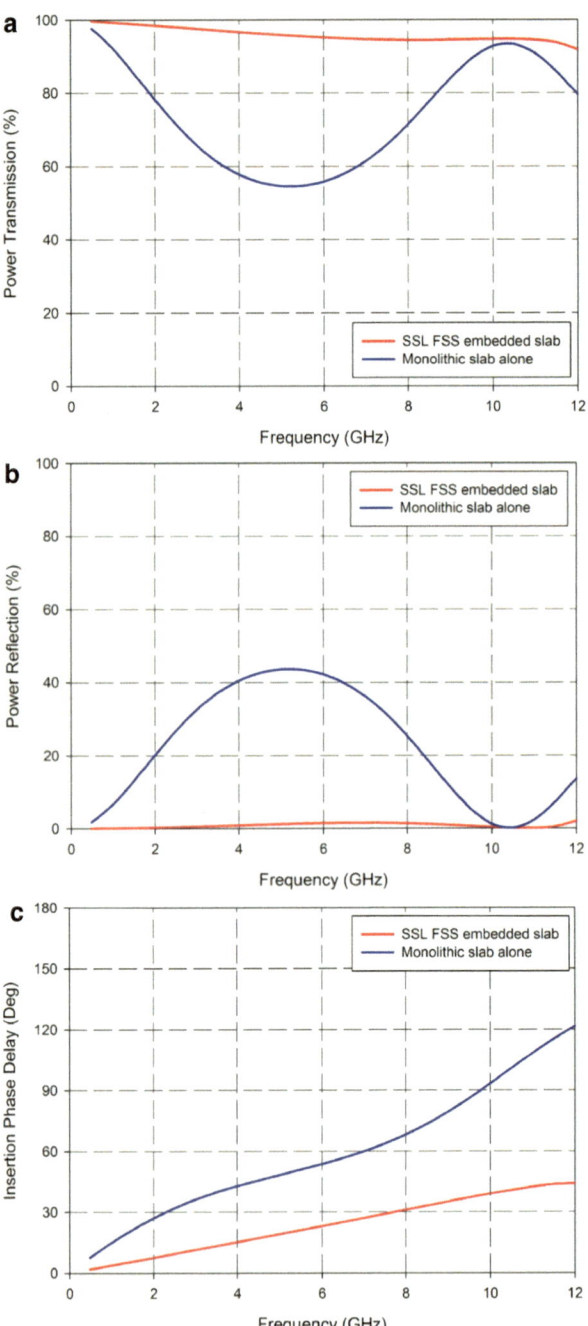

Fig. 5.4 **a** Transmission efficiency of SSL-FSS embedded radome and monolithic radome at 30° over a broadband (0.5–12 GHz). **b** Reflection characteristics of SSL-FSS embedded radome and monolithic radome at 30° over a broadband (0.5–12 GHz). **c** Insertion phase delay of SSL-FSS embedded radome and monolithic radome at 30° over a broadband (0.5–12 GHz)

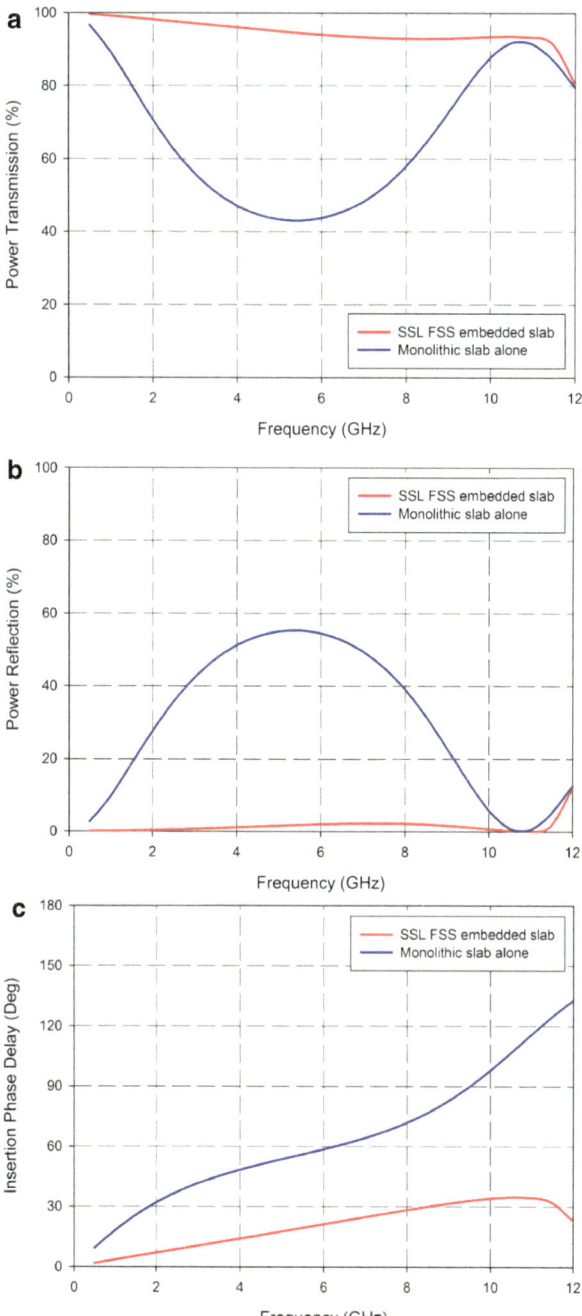

Fig. 5.5 **a** Transmission efficiency of SSL-FSS embedded radome and monolithic radome at 45° over a broadband (0.5–12 GHz). **b** Reflection characteristics of SSL-FSS embedded radome and monolithic radome at 45° incidence angle over a broadband (0.5–12 GHz). **c** Insertion phase delay of SSL-FSS embedded radome and monolithic radome at 45° over a broadband (0.5–12 GHz)

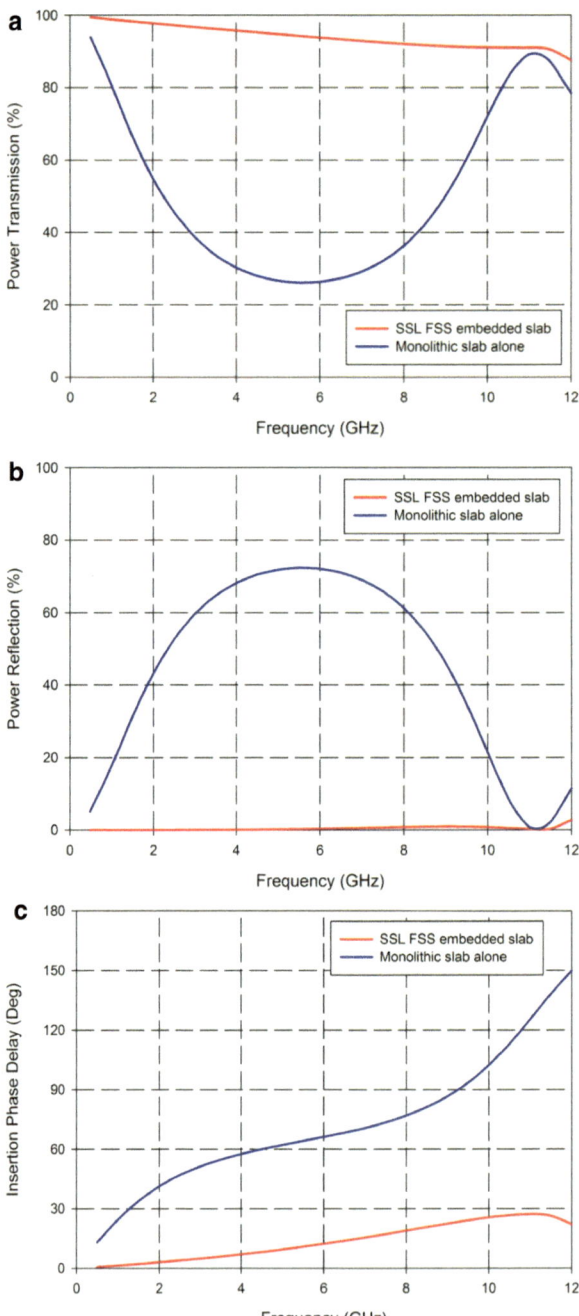

Fig. 5.6 a Transmission efficiency of SSL-FSS embedded radome and monolithic radome at 60° over a broadband (0.5–12 GHz). **b** Reflection characteristics of SSL-FSS embedded radome and monolithic radome at 60° over a broadband (0.5–12 GHz). **c** Insertion phase delay of SSL-FSS embedded radome and monolithic radome at 60° over a broadband (0.5–12 GHz)

further as compared to lower incidence angle cases. In the case of power reflection also, the value has increased at higher frequencies. But at the same time, the result is better than that for the monolithic radome wall alone case. Similarly, the IPD offered by the SSL-FSS embedded radome wall is also far better than that of monolithic radome wall alone case at incidence angle 60°. From the IPD study, BSE can also be computed, and SSL-FSS provides better results than that of monolithic radome wall alone structure.

5.1.4 Conclusion

Monolithic radome walls exhibit very narrow bandwidth. In this study, the EM analysis of SSL-FSS embedded radome wall offers better performance over a broadband of frequencies at various incidence angles as compared to monolithic radome walls which is requisite in modern radar antenna systems like active electronically scanned arrays. Furthermore, inclusion of FSS embedded structures offers extended structural rigidity to monolithic radome walls.

5.2 Broadbanding Techniques Based on Double Square Loop Frequency Selective Surfaces (DSL-FSS)

Various band stop filters and reflector antennas make extensive use of double square loop (DSL) elements in common. Likewise, for multiplexing in the S, X, and Ku bands, DSL-FSS employed sub-reflectors have been used (Wu 1992).

C-sandwich radome wall configuration embedded with DSL-FSS structures and the EM performance study over a frequency range of 1–15 GHz is presented in this section.

5.2.1 EM Design Considerations of DSL-FSS Embedded C-sandwich Radome Wall Structures

A C-sandwich radome wall is considered to be two A-sandwiched walls connected back-to-back, i.e., it consists of inner and outer skins and a middle layer of same material (here, glass epoxy: $\varepsilon_r = 4.0$; $\tan\delta_e = 0.015$, inner and outer skin thickness: 0.75 mm, middle layer: 1.5 mm). Inner and outer cores of same material ($\varepsilon_r = 1.15$; $\tan\delta_e = 0.005$) are sandwiched between inner and outer skin layers and middle layer, respectively. Thickness of C-sandwich inner and outer cores is optimized for maximum power transmission over the frequency band 1–15 GHz is 5.44 mm. DSL-FSS structure is shown in Fig. 5.7a. DSL-FSS structures are embedded in the middle of the inner and outer cores, respectively, as shown in Fig. 5.7b.

After optimization of the design parameters such as length, width, and gap of the DSL-FSS structure at different angle of incidence, it is perceived that DSL-FSS provides better performance parameters at lower incidence angles such as 0° and 45° and the parameters for these particular angles are listed in Table 5.2.

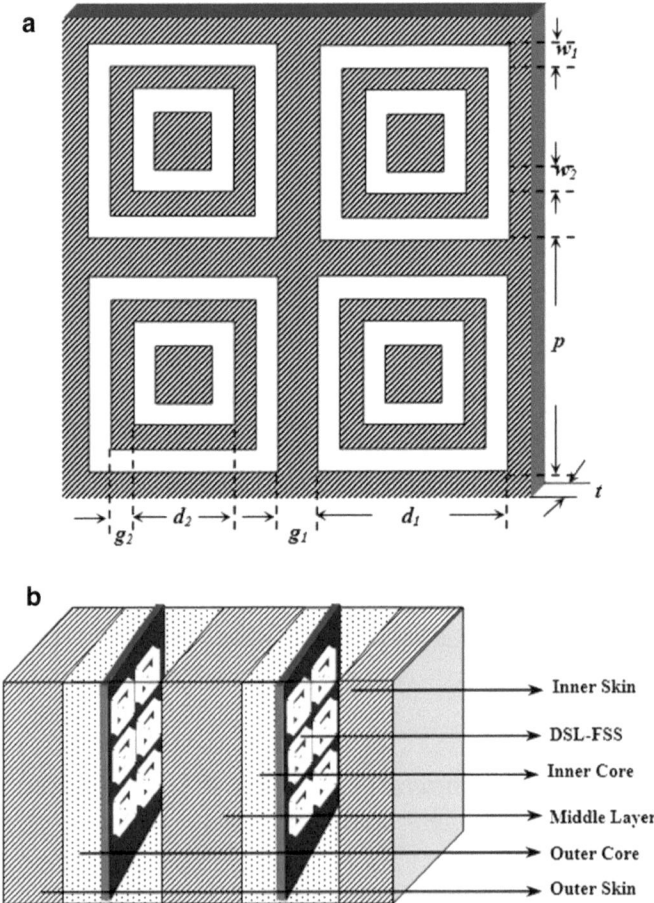

Fig. 5.7 **a** Schematic of DSL-FSS. **b** Schematic of DSL-FSS embedded C-sandwich radome panel

Table 5.2 Design parameters of DSL-FSS

Design parameters (mm)	Angle of incidence	
	0°	45°
Inner square length, d_2	3.0	3.0
Inner square width, w_2	1.5	1.5
Inner gap, g_2	1.5	1.5
Outer square length, d_1	9.0	9.0
Outer square width, w_1	1.5	1.5
Outer gap, g_1	3.9	0.6
Pitch, p	12.9	9.6
Thickness of FSS	1.8	1.3

5.2.2 EM Performance Analysis of Double Square Loop FSS Embedded Monolithic Radome

The DSL-FSS is placed along the mid-plane of C-sandwich cores.

As discussed in Chap. 2, equivalent transmission line method is used to determine the EM performance parameters of DSL-FSS embedded C-sandwich structures, whereas different sections in the transmission line represent corresponding layer of the DSL-FSS embedded C-sandwich structure.

As used in the previous sections, let Z_0 be the free space characteristic impedance and that of FSS, core and skin are given by Z_s, Z_c, and Z_{FSS}, respectively. Similarly, electrical length of each layer is represented by Φ.

In the equivalent transmission line method, the matrix consisting of A_i, B_i, C_i, and D_i parameters will represent the ith layer of the radome wall. The thicknesses of skin and middle layers are given by t_s and $2t_s$, respectively.

Let t_c represent the thickness of core and d_{FSS} that of DSL-FSS layer, respectively.

C-sandwich wall alone structure matrix representation is shown below. The outer skin of the DSL-FSS embedded C-sandwiched radome wall is represented by

$$\begin{bmatrix} A_1 & B_1 \\ C_1 & D_1 \end{bmatrix} = \begin{bmatrix} \cos \Phi_1 & j\frac{z_s}{z_0} \sin \Phi_1 \\ j\frac{z_0}{z_s} \sin \Phi_1 & \cos \Phi_1 \end{bmatrix} \qquad (5.8)$$

Similarly, the outer core section is also represented by

$$\begin{bmatrix} A_2 & B_2 \\ C_2 & D_2 \end{bmatrix} = \begin{bmatrix} \cos \Phi_2 & j\frac{z_c}{z_0} \sin \Phi_2 \\ j\frac{z_0}{z_c} \sin \Phi_2 & \cos \Phi_2 \end{bmatrix} \qquad (5.9)$$

DSL-FSS layer is represented by the ABCD matrix with parameters A_{FSS}, B_{FSS}, C_{FSS}, and D_{FSS} and is given by

$$\begin{bmatrix} A_{FSS} & B_{FSS} \\ C_{FSS} & D_{FSS} \end{bmatrix} = \begin{bmatrix} 1 & 0 \\ 1/jX_r & 1 \end{bmatrix} \tag{5.10}$$

In the above equation, X_r is the reactance of DSL-FSS structure and is a function of inductive reactance X and capacitive susceptance B. According to equivalent circuit design, (Langley and Parker 1983) of SSL-FSS,

For TE incidence, X and B are given by

$$X_{TE} = \frac{p\cos\theta}{\lambda}\left[\ln\left(\frac{1}{\sin(\pi\omega/2p)}\right)\right] \tag{5.11}$$

$$B_{TE} = \frac{4p\sec\theta}{\lambda}\left[\ln\left(\frac{1}{\sin(\pi g/2p)}\right)\right] \tag{5.12}$$

In the above equations, g, p, and w represent gap, periodicity, and width of DSL-FSS, respectively. λ represents the wavelength of the incident EM wave.

The medium present in the aperture sections of the DSL-FSS is assumed to be the same as that of the core, to reduce the impedance mismatch.

The middle layer:

$$\begin{bmatrix} A_3 & B_3 \\ C_3 & D_3 \end{bmatrix} = \begin{bmatrix} \cos\Phi_3 & j\frac{z_s}{z_0}\sin\Phi_3 \\ j\frac{z_0}{z_s}\sin\Phi_3 & \cos\Phi_3 \end{bmatrix} \tag{5.13}$$

Inner core is expressed as

$$\begin{bmatrix} A_4 & B_4 \\ C_4 & D_4 \end{bmatrix} = \begin{bmatrix} \cos\Phi_4 & j\frac{z_c}{z_0}\sin\Phi_4 \\ j\frac{z_0}{z_c}\sin\Phi_4 & \cos\Phi_4 \end{bmatrix} \tag{5.14}$$

The inner skin:

$$\begin{bmatrix} A_5 & B_5 \\ C_5 & D_5 \end{bmatrix} = \begin{bmatrix} \cos\Phi_5 & j\frac{z_s}{z_0}\sin\Phi_5 \\ j\frac{z_0}{z_s}\sin\Phi_5 & \cos\Phi_5 \end{bmatrix} \tag{5.15}$$

Thus, the DSL-FSS embedded C-sandwich radome wall is represented by

$$\begin{bmatrix} A & B \\ C & D \end{bmatrix} = \begin{bmatrix} A_1 & B_1 \\ C_1 & D_1 \end{bmatrix}\begin{bmatrix} A_2 & B_2 \\ C_2 & D_2 \end{bmatrix}\begin{bmatrix} 1 & 0 \\ 1/jX_r & 1 \end{bmatrix}\begin{bmatrix} A_2 & B_2 \\ C_2 & D_2 \end{bmatrix}$$
$$\begin{bmatrix} A_3 & B_3 \\ C_3 & D_3 \end{bmatrix}\begin{bmatrix} A_4 & B_4 \\ C_4 & D_4 \end{bmatrix}\begin{bmatrix} 1 & 0 \\ 1/jX_r & 1 \end{bmatrix}\begin{bmatrix} A_4 & B_4 \\ C_4 & D_4 \end{bmatrix}\begin{bmatrix} A_5 & B_5 \\ C_5 & D_5 \end{bmatrix} \tag{5.16}$$

Using Eq. (5.16), it represents the final ABCD matrix of the whole structure. The power transmission coefficient and power reflection coefficient are given by Eqs. (3.10) and (3.11).

The IPD is given by

$$\text{IPD} = -\angle T - \frac{2\pi}{\lambda}(4t_{\text{s}} + 4t_{\text{c}} + 2d_{\text{FSS}})\cos\theta \tag{5.17}$$

In the above expression, $\angle T$ is the phase angle of the corresponding voltage transmission coefficient T of the DSL-FSS embedded radome wall structure.

5.2.3 Numerical Results and Discussion

Based on the equivalent transmission line model, the EM performance parameters of DSL-FSS embedded C-sandwich radome wall are assessed over a wide frequency band ranging from 1 to 15 GHz for different angles of incidence such as 0°, 30°, and 45°, respectively.

Power transmission characteristics of C-sandwich alone radome wall structure at normal angle of incidence are illustrated in Fig. 5.8a. The performance degrades drastically from 11 to 15 GHz. DSL-FSS embedded C-sandwiched radome wall shows better performance than that of C-sandwich wall alone structure. Figure 5.8b, c illustrates the power reflection and IPD characteristics of the radome wall structures at normal angles of incidence, respectively. Power reflection characteristics of DSL-FSS embedded radome wall structure are considerably less than that of C-sandwich alone radome walls structure. Likewise, at normal incidence, DSL-FSS embedded C-sandwich radome wall structure offers better IPD characteristics than that of C-sandwich alone radome wall configuration.

Figure 5.9a–c illustrates the EM performance parameters of C-sandwich wall alone and DSL-FSS embedded C-sandwich radome walls at an incidence angle of 30°. In this case also, DSL-FSS embedded C-sandwich radome walls exhibit better EM performance characteristics as compared to C-sandwich alone case. But as the incidence angle is increased to 30°, the performance curves slightly degrade as compared to that of normal incidence. Still, the shape of the curves remains almost similar as in the former case.

If the incidence angle is further increased, such as beyond 45°, the EM performance parameters of the C-sandwich radome wall alone deteriorate drastically and do not fulfill the criteria for radome wall applications.

For an incidence angle of 45°, the performance parameters of DSL-FSS embedded C-sandwich radome wall and C-sandwich radome wall alone structure are shown in Fig. 5.10a–c.

DSL-FSS embedded C-sandwich radome wall structure offers low power reflection, IPD, and gives better performance.

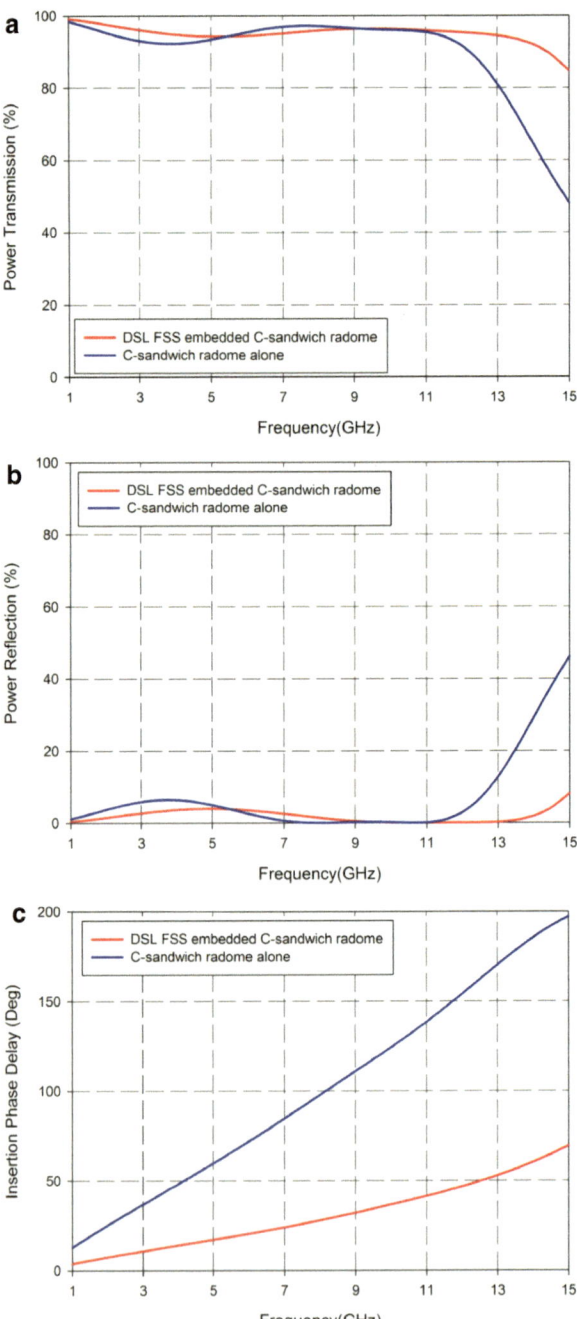

Fig. 5.8 **a** Transmission efficiency of DSL-FSS embedded C-sandwich radome and C-sandwich radome alone at normal incidence. **b** Reflection characteristics of DSL-FSS embedded C-sandwich radome and C-sandwich radome alone at normal incidence. **c** Insertion phase delay of DSL-FSS embedded C-sandwich radome and C-sandwich radome alone at normal incidence

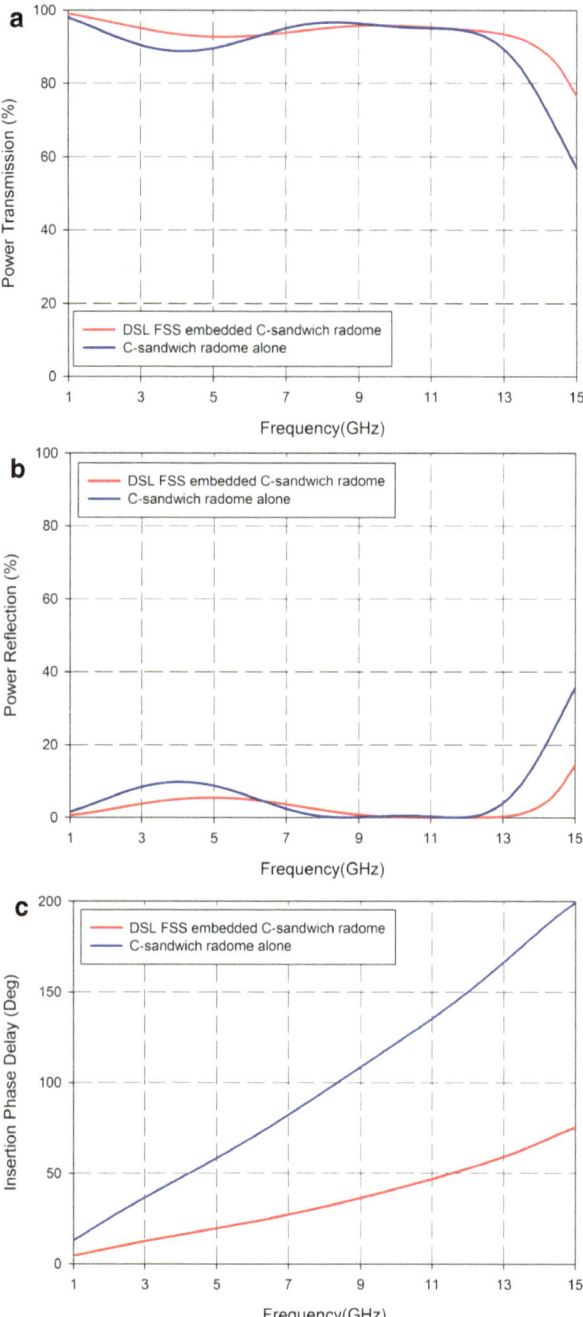

Fig. 5.9 **a** Transmission efficiency of DSL-FSS embedded C-sandwich radome and C-sandwich radome alone at 30°. **b** Reflection characteristics of DSL-FSS embedded C-sandwich radome and C-sandwich radome alone at 30°. **c** Insertion phase delay of DSL-FSS embedded C-sandwich radome and C-sandwich radome alone at 30°

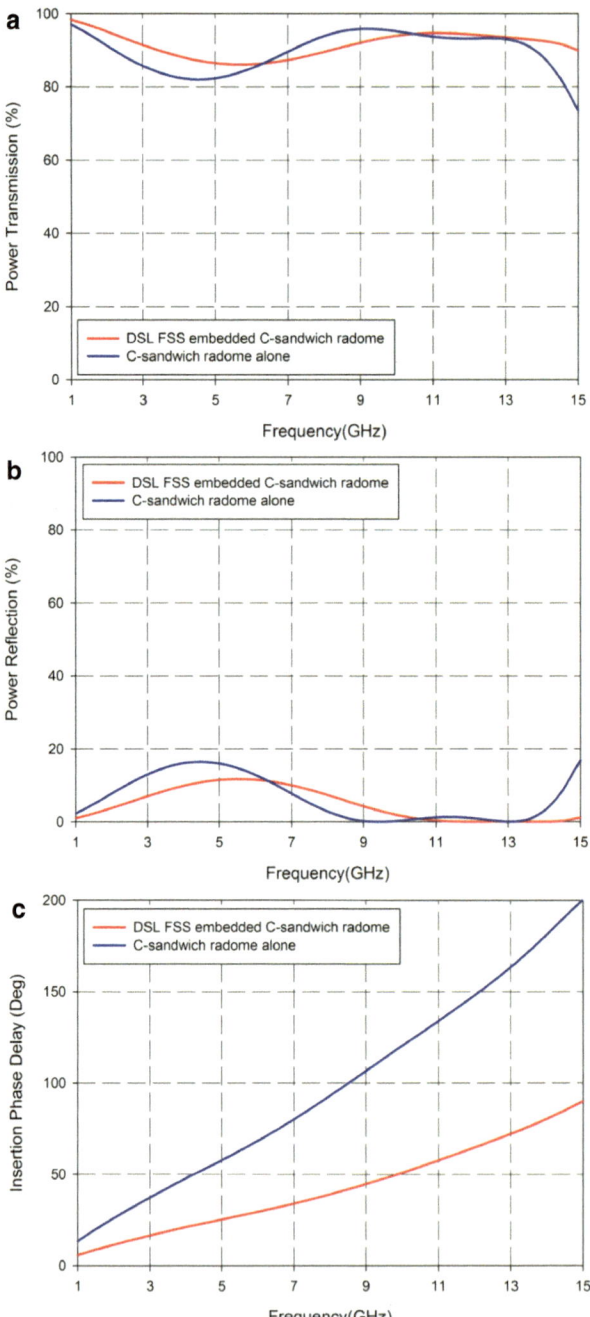

Fig. 5.10 **a** Transmission efficiency of DSL-FSS embedded C-sandwich radome and C-sandwich radome alone at 45°. **b** Reflection characteristics of DSL-FSS embedded C-sandwich radome and C-sandwich radome alone at 45°. **c** Insertion phase delay of DSL-FSS embedded C-sandwich radome and C-sandwich radome alone at 45°

5.2.4 Conclusion

In this section, design and analysis of EM performance of C-sandwich alone radome wall structures and DSL-FSS embedded radome wall structures are performed. In the analysis, it is found that the DSL-FSS embedded radome wall structure offers better EM performance over a wide band of frequencies, including L, S, C, and X bands. It is also observed that at high incidence angles, such as 45°, the performance deteriorates and thus the embedded structure is not suitable for high incidence angle applications.

5.3 Broadbanding Techniques Based on Jerusalem Cross-Frequency Selective Surfaces (JC- FSS)

EM performance improvement of A-sandwich radome using frequency selective surfaces is presented in this section. Jerusalem cross-frequency selective surface (JC-FSS) array is used for enhancing the performance of the radome which is embedded in mid-plane of the core of A-sandwich radome. The frequency band considered here is 1–12 GHz. The equivalent transmission line method is used for the estimation of EM performance parameters as explained in Chap. 2. An equivalent transmission line model can be derived from the Jerusalem cross FSS embedded A-sandwich radome with different sections corresponding to skin, core, and JC-FSS. A comparative investigation of JC-FSS included A-sandwich radome and A-sandwich radome of identical material and thickness (core and skin layers) is done in this section. The FSS structure embedded radome gives an excellent EM performance over the conventional radome wall configuration which makes it a desirable choice for the design of normal incidence radomes (hemispherical/cylindrical), near-normal incidence radomes (paraboloidal), and highly streamlined airborne nosecone radomes.

Several techniques for improving the EM performance of radomes have been reported (Cary 1983; Kozakoff 2010). Monolithic radomes with metallic inclusions loaded in the core improve the bandwidth of the radome to a greater extent (Frenkel 2001). For superior EM performance over a broadband frequency range, minimal boresight error, high transmission efficiency, and low cross-polarization levels are required.

For superior EM performance applications, frequency selective surface (FSS)-based radome wall structures are desirable (Munk 2005; Lin et al. 2009). Recently, metamaterial-based FSS are widely used in the design of novel radome configurations (Latrach et al. 2010; Basiry et al. 2011; Narayan et al. 2012).

Due to high structural rigidity, A-sandwich radomes are generally used in airborne applications as compared to monolithic wall configurations (Kozakoff 2010). At high incidence angles, the EM performance of the A-sandwich radomes degrades. Application of Jerusalem cross FSS structures for enhancing the EM performance parameters over a broadband of frequencies (1–12 GHz) for different angles of incidence of the radome wall is discussed in this section. For this study, an A-sandwich model has been loaded with an array of Jerusalem cross structures in the mid-plane and studied the EM performance of the new structure. Jerusalem cross embedded A-sandwich radome panel shows better performance over A-sandwich alone structures.

5.3.1   EM Design Aspects of JC-FSS Embedded A-sandwich Radome

Unit cell of Jerusalem cross structure is as shown in Fig. 5.11a and that of aperture-type Jerusalem cross structure is as shown in Fig. 5.11b. Practically, Jerusalem cross FSS structures are made by removing the corresponding sections of crossed dipoles and end-loading elements from the thick metallic sheets made up of copper or silver and embedding this in the mid-plane of A-sandwich radomes to form Jerusalem cross FSS embedded A-sandwich radome as shown in Fig. 5.11c. Glass epoxy (ε_r = 4.0 and $\tan\delta_e$ = 0.015) with thickness 0.75 mm and foam core (ε_r = 1.15 and $\tan\delta_e$ = 0.002) with thickness 5.44 mm is used for this radome panel to operate from 1 to 12 GHz. Electric fields along the inner edges of the aperture-type Jerusalem cross (vertical inductive and horizontal capacitive sections) are the major sources that contribute to the EM performance of the whole structure.

Equivalent circuit method is used to design the Jerusalem cross grid inductive susceptance, and equivalent transmission line method is used to model the Jerusalem cross embedded monolithic A-sandwich radome wall. Length—d, gap width—g, pitch—p, thickness—t, and element width—w are the design parameters of Jerusalem cross FSS unit cell. They are optimized to obtain better EM performance parameters in the band 1–12 GHz. Table 5.3 describes the optimized parameters for perpendicular polarization at incidence angles 0°, 45°, and 80°.

Equivalent transmission line method (Cary 1983) in conjunction with equivalent circuit model (Anderson 1975; Ohira 2005) is used to compute the EM performance parameters of the Jerusalem cross embedded A-sandwich radome wall configuration.

5.3.2   EM Performance Analysis of JC-FSS Embedded A-sandwich Radome

According to equivalent transmission line method, A-sandwich radome wall, considered as low impedance lines connected end to end is represented by a single matrix with A, B, C, and D parameters, which is a result of multiplication of individual

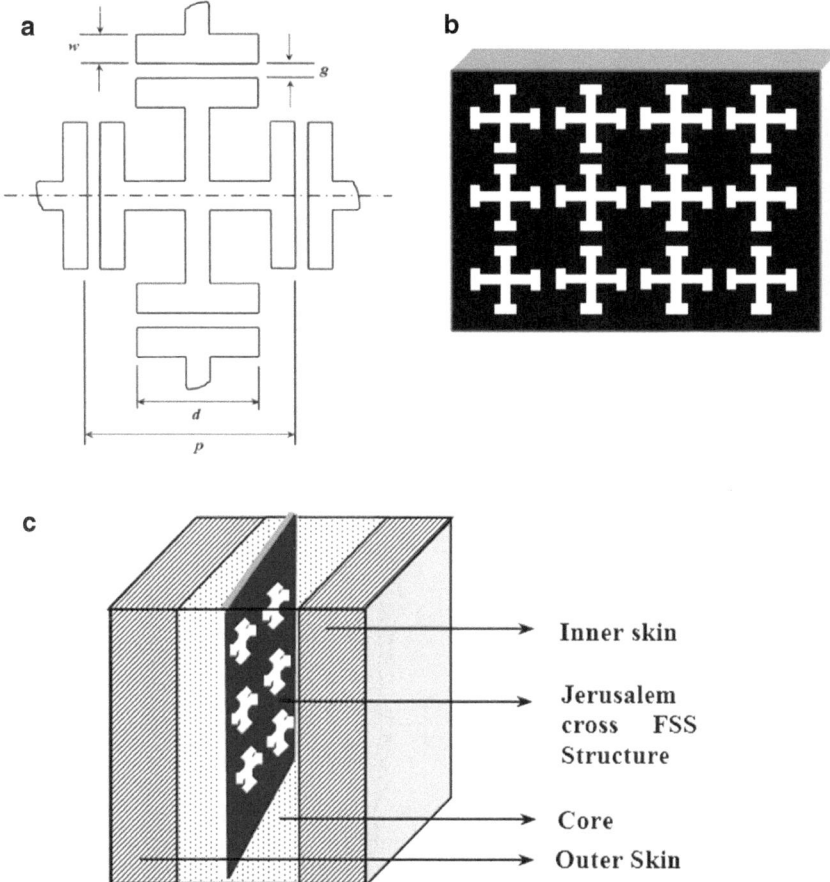

Fig. 5.11 **a** Unit cell of Jerusalem cross FSS. **b** Jerusalem cross FSS (aperture type). **c** Schematic of A-sandwich radome centrally loaded with aperture-type Jerusalem cross FSS

Table 5.3 Design parameters of Jerusalem cross FSS

Angle of incidence	Width, w (mm)	Length, d (mm)	Gap width, g (mm)	Pitch, p (mm)	Thickness of FSS, t (mm)
0°	0.50	1.1	0.55	2.21	1.35
45°	1.00	4.50	2.40	11.40	1.55
80°	1.80	12.5	1.40	22.60	1.60

matrices corresponding to each layer of the radome wall. ith layer of the radome wall is represented by a matrix with elements A_i, B_i, C_i, and D_i.

Each layer of the Jerusalem cross FSS embedded A-sandwich radome wall is represented as matrices are as shown below.

The outer skin of the radome wall:

$$\begin{bmatrix} A_1 & B_1 \\ C_1 & D_1 \end{bmatrix} = \begin{bmatrix} \cos \Phi_1 & j\frac{Z_s}{Z_0} \sin \Phi_1 \\ j\frac{Z_0}{Z_s} \sin \Phi_1 & \cos \Phi_1 \end{bmatrix} \tag{5.18}$$

In the above expression, characteristic impedance of free space is represented by Z_0. Φ represents the electrical length corresponding to each layer (it is a function of incident angle$_\theta$), dielectric layer thickness d, wavelength of incident wave, and dielectric layer complex permittivity.

Z_s, Z_c, and Z_{FSS} represent the characteristic impedances of skin, core and that of Jerusalem cross FSS.

A-sandwich core is embedded with Jerusalem cross FSS in the mid-plane and thus the core is considered as two symmetrical halves.

First symmetric half-section of FSS embedded radome wall core is shown as

$$\begin{bmatrix} A_2 & B_2 \\ C_2 & D_2 \end{bmatrix} = \begin{bmatrix} \cos \Phi_2 & j\frac{Z_c}{Z_0} \sin \Phi_2 \\ j\frac{Z_0}{Z_c} \sin \Phi_2 & \cos \Phi_2 \end{bmatrix} \tag{5.19}$$

Matrix representing Jerusalem cross FSS with parameters A_{FSS}, B_{FSS}, C_{FSS}, and D_{FSS} is given as

$$\begin{bmatrix} A_{FSS} & B_{FSS} \\ C_{FSS} & D_{FSS} \end{bmatrix} = \begin{bmatrix} 1 & 0 \\ \frac{1}{jX_r} & 1 \end{bmatrix} \tag{5.20}$$

In the above expression, X_r is the reactance of Jerusalem cross FSS. Inductive reactance X and capacitive susceptance B contribute to X_r (Anderson 1975).

X and B are represented as following, from equivalent transmission line method.

$$X = \frac{p}{\lambda}\left[\ln\left(\cosec\left(\frac{\pi w}{2 * p}\right)\right) + F\right] \tag{5.21}$$

$$B = \frac{4d}{\lambda}\left\{\ln\left(\cosec\left(\frac{\pi g}{2p}\right)\right) + F + \frac{\pi t}{2g}\right\} \tag{5.22}$$

where F is the correction factor given by

$$F = \frac{Q \, c^2}{1 + Q \, s^2} + \left[\frac{pc}{4\lambda}(1 - 3s)\right]^2 \tag{5.23}$$

Q, c ad s are the coefficients and is given by

$$Q = \left[1 - \left(\frac{p}{\lambda}\right)^2\right]^{-\frac{1}{2}} - 1 \tag{5.24}$$

where $c = \cos^2 \frac{\pi w}{2p}$; and $s = 1 - c$. The second half-section of the core is represented by

$$\begin{bmatrix} A_3 & B_3 \\ C_3 & D_3 \end{bmatrix} = \begin{bmatrix} \cos \Phi_4 & j\frac{z_c}{z_0} \sin \Phi_4 \\ j\frac{z_0}{z_c} \sin \Phi_4 & \cos \Phi_4 \end{bmatrix} \tag{5.25}$$

The inner skin is represented by

$$\begin{bmatrix} A_4 & B_4 \\ C_4 & D_4 \end{bmatrix} = \begin{bmatrix} \cos \Phi_4 & j\frac{z_c}{z_0} \sin \Phi_4 \\ j\frac{z_0}{z_c} \sin \Phi_4 & \cos \Phi_4 \end{bmatrix} \tag{5.26}$$

Single matrix corresponding to the whole radome wall configuration is given as

$$\begin{bmatrix} A & B \\ C & D \end{bmatrix} = \begin{bmatrix} A_1 & B_1 \\ C_1 & D_1 \end{bmatrix} \begin{bmatrix} A_2 & B_2 \\ C_2 & D_2 \end{bmatrix} \begin{bmatrix} 1 & 0 \\ \frac{1}{jX_r} & 1 \end{bmatrix} \begin{bmatrix} A_3 & B_3 \\ C_3 & D_3 \end{bmatrix} \begin{bmatrix} A_4 & B_4 \\ C_4 & D_4 \end{bmatrix} \tag{5.27}$$

Equation (3.1) is used to find out the A, B, C, and D parameters of the resultant matrix. Equations (3.10) and (3.11) give the expression for power transmission and reflection coefficients, respectively.

IPD is the measure of phase distortion of the EM wave incurred by the radome wall. IPD of the Jerusalem cross embedded A-sandwich radome wall is given by

$$\text{IPD} = -\angle T - \frac{2\pi}{\lambda}(2t_s + 2t_c + d_{\text{FSS}}) \cos \theta \tag{5.28}$$

Which includes IPD due to inner as well as outer skin layers, core layers, and FSS layer, where t_s and t_c represent the thicknesses of skin as well as core the radome. $\angle T$ represents the phase angle related to the voltage transmission coefficient of the whole structure.

5.3.3 Numerical Results and Discussion

At various angles of incidence such as normal incidence, 45° and 80°, the EM performance of the Jerusalem cross FSS embedded A-sandwich radome wall structures is studied for perpendicular polarization. From Fig. 5.12a, b, it is evident that the Jerusalem cross FSS embedded A-sandwich radome wall shows superior EM performance over A-sandwich alone radome wall structures over the frequency band 1–12 GHz.

At normal angle of incidence, Jerusalem cross FSS embedded A-sandwich radome wall has much lower IPD as compared to A-sandwich radome wall alone radome wall structures, as depicted in Fig. 5.12c.

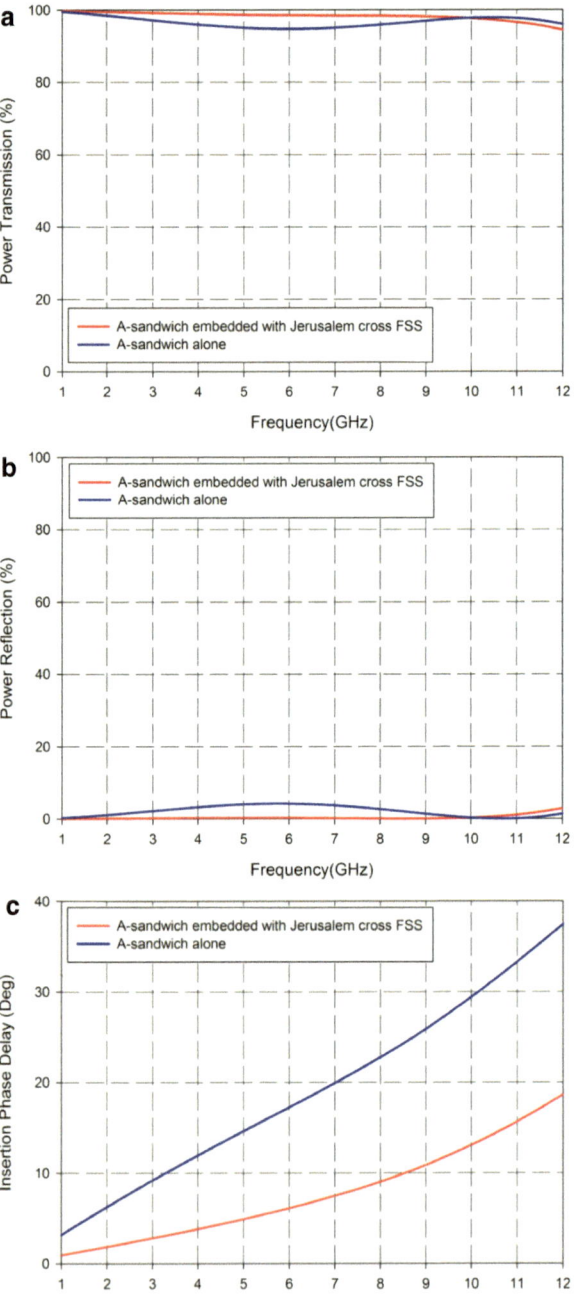

Figure 5.13a–c depicts the EM performance parameters at an incidence angle of 45°. In this case also, Jerusalem cross FSS embedded A-sandwich radome wall shows superior EM performance than that of A-sandwich alone radome wall configuration. Jerusalem cross FSS offers power transmission efficiency better than 95% as noticeable from Fig. 5.13a. Similarly, Jerusalem cross FSS embedded A-sandwich radome wall structure shows power reflection lower than 5%, and it is much better than that of A-sandwich alone radome wall structure. It is represented in Fig. 5.13b. Likewise, Fig. 5.13c compares the IPD of A-sandwich radome wall alone and FSS embedded radome wall structures. From the above study, it is clear that the FSS embedded A-sandwich radome wall gives better performance at 45° as compared to normal incidence case.

EM performance parameters of FSS embedded radome at a higher incidence angle of 80° are also studied and compare with that of A-sandwich alone radome wall structure in Fig. 5.14a–c. In this case also, power transmission efficiency of FSS embedded radomes is much higher than that of A-sandwich alone radome wall, and all other parameters such as power reflection and IPD are lower than that of A-sandwich alone case.

It is observed from the study that the Jerusalem cross embedded A-sandwich radome wall shows better EM performance than that of A-sandwich alone radome wall structures at normal incidence as well as 45° and 80° incidence angles.

5.3.4   Conclusion

Inclusion of Jerusalem cross FSS to enhance the EM performance of A-sandwich radome walls has been discussed in this section. From the study, it is observed that the Jerusalem cross FSS embedded A-sandwich radome walls exhibit superior EM performance parameters over the frequency band 1–12 GHz at various angles of incidence such as at normal incidence, 45° and 80°, respectively. Among that, normal incidence and 45° incidence angles show better EM performance. For larger scan angles, Jerusalem cross FSS embedded A-sandwich radome walls are highly recommended over A-sandwich alone radome wall structures.

Fig. 5.13 a Transmission efficiency of Jerusalem cross FSS embedded A-sandwich radome and A-sandwich radome alone at 45°. **b** Reflection characteristics of Jerusalem cross FSS embedded A-sandwich radome and A-sandwich radome alone at 45°. **c** Insertion phase delay of Jerusalem cross FSS embedded A-sandwich radome and A-sandwich radome alone at 45°

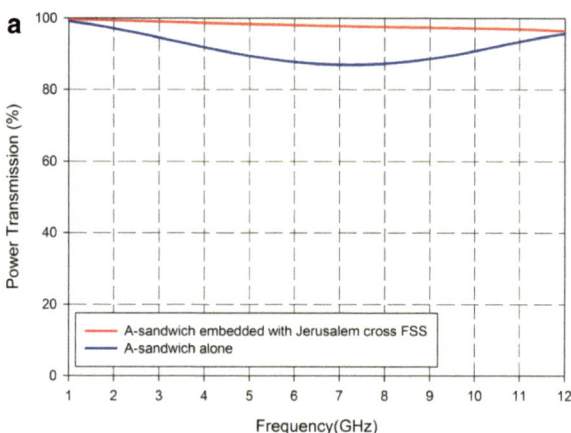

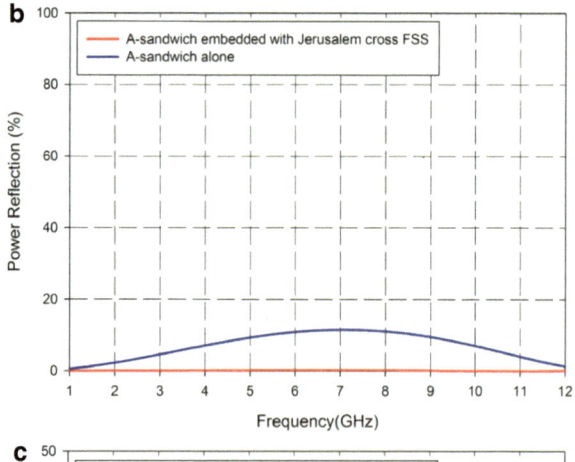

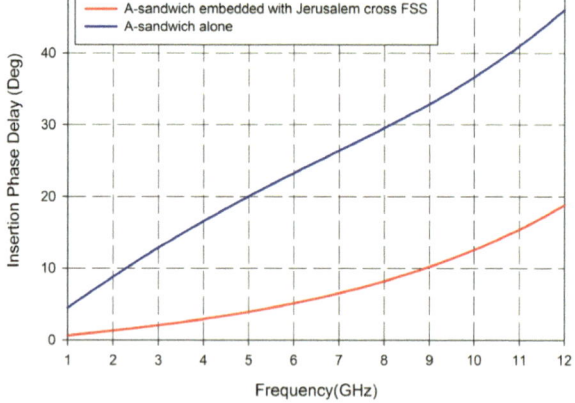

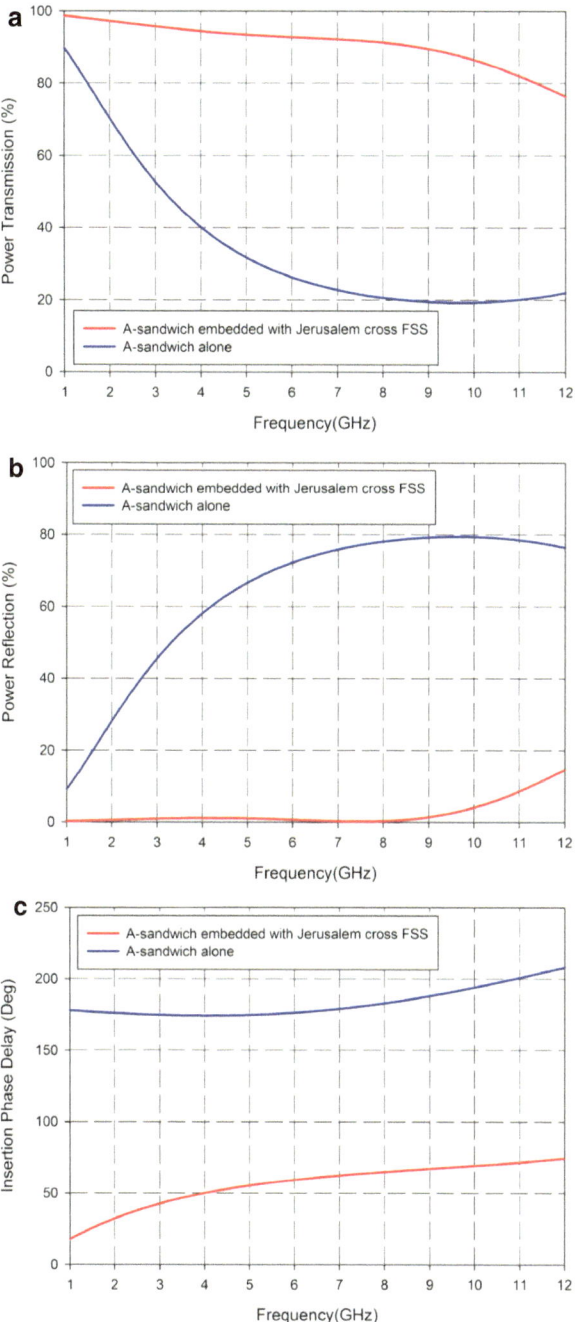

Fig. 5.14 a Transmission efficiency of Jerusalem cross FSS embedded A-sandwich radome and A-sandwich radome alone at 80°. **b** Reflection characteristics of Jerusalem cross FSS embedded A-sandwich radome and A-sandwich radome alone at 80°. **c** Insertion phase delay of Jerusalem cross FSS embedded A-sandwich radome and A-sandwich radome alone at 80°

References

Anderson I (1975) On the theory of self-resonant grids. Bell Syst Tech J 54:1725–1731

Basiry R, Abiri H, Yahaghi A (2011) Electromagnetic performance analysis of omega type metamaterial radome. Int J RF Microwave Comput Aided Eng 21(6):665–673

Cary RHJ (1983) Radomes. In: Rudge AW, Milne K, Olver AD, Knight P (eds.) The handbook of antenna design. Peter Peregrinus, London, UK. ISBN 0-906048-82-6, 960 p

Frenkel A (2001) Thick metal dielectric radome. IET Electron Lett 38:1374–1375

Kozakoff DJ (2010) Analysis of radome enclosed antennas. Artech House, Norwood. ISBN 1596934417, 294 p

Langley RJ, Parker EA (1983) Double square frequency selective surface and their equivalent circuit. IET Electron Lett 19(17):675–677

Latrach M, Rmili H, Sabatier C, Seguenot E, Toutain S (2010) Design of a new type of metamaterial radome at low frequencies. Microw Opt Technol Lett 52(5):1119–1123

Lin BQ, Li F, Zheng QR, Zen YS (2009) Design and simulation of a miniature thick-screen frequency selective surface radome. Prog Electromag Res 138:537–553

Munk BA (2005) Frequency selective surfaces—theory and design. Wiley, New York

Narayan S, Shamala JB, Nair RU, Jha RM (2012) Electromagnetic performance analysis of novel multiband metamaterial FSS for millimeter wave radome applications. Comput Mater Cont 31(1):1–16

Ohira MH, Deguchi MT, Shigesawa H (2005) Analysis of frequency selective surface with arbitrary shaped element by equivalent circuit model. Electron Commun Jpn Part 2 88:9–17

Vardaxoglou JC (1997) Frequency selective surfaces—analysis and design. Research Studies Press, Taunton

Walton JD (1966) Techniques for airborne radome design, AFAL report, no. 45433. Wright-Patterson AFB, pp 105–108

Wu TK (1992) Single-screen triband FSS with double-square-loop elements. Microwav Opt Tech Lett 5:56–59

Chapter 6
Broadbanding Techniques Based on Inhomogeneous Dielectric Structures

Emergence of inhomogeneous planar layer (IPL) was a major breakthrough in the field of aerospace domain. IPL is widely used in the design of shields, filters, and radomes (Toscano et al. 2001; Venkatarayalu et al. 2005; Amirhosseini 2006). IPL provides an exceptional impedance matching. Therefore, it is an inevitable choice for the design and development of radomes. Monolithic half-wave radome wall and A-sandwich radome wall are generally used in the airborne radome applications because of its superior EM performance at the design frequency. Its applications are limited due to narrow bandwidth. Hence, for the enhancement of EM performance parameters over the entire broadband of frequencies at normal incidence as well as at high incidence angles, an IPL formed by the exponential variation of the complex permittivity incorporated in the single layer radome wall is presented in this work.

In practice, IPL radome can be realized by stacking the layers of differing complex permittivity. Similarly, a required dielectric gradient within the symmetric half-slab can be achieved by vacuum bagging by taking vacuum pressure and time as the control parameters. In this chapter, EM performance parameters of the IPL radome are compared with that of half-wave radome of identical thickness to substantiate the superior EM performance of IPL radome. IPL-based A-sandwich wall configuration is obtained by incorporating exponential variation of relative complex permittivity in the core as shown in Fig. 6.1b.

6.1 EM Design Aspects of Inhomogeneous Planar Layer (IPL) Radome and A-sandwich Radome with IPL Core

IPL Radome: In order to attain maximum power transmission at 10 GHz, the optimized thickness of monolithic half-wave radome wall (glass epoxy: $\varepsilon_r = 4$, $\tan\delta_e = 0.015$) is 7.44 mm. Here the IPL radome has exponential variation of complex permittivity, and its thickness is same as shown in Fig. 6.1a.

© The Author(s), under exclusive license to Springer Nature Singapore Pte Ltd. 2020   61
P. S. Mohammed Yazeen et al., *Broadbanding Techniques for Radomes*,
SpringerBriefs in Computational Electromagnetics,
https://doi.org/10.1007/978-981-33-4130-2_6

Fig. 6.1 a Schematic of IPL radome (Monolithic) wall with exponential variation of relative permittivity. **b** Exponential variation of relative permittivity across the core of A-sandwich radome

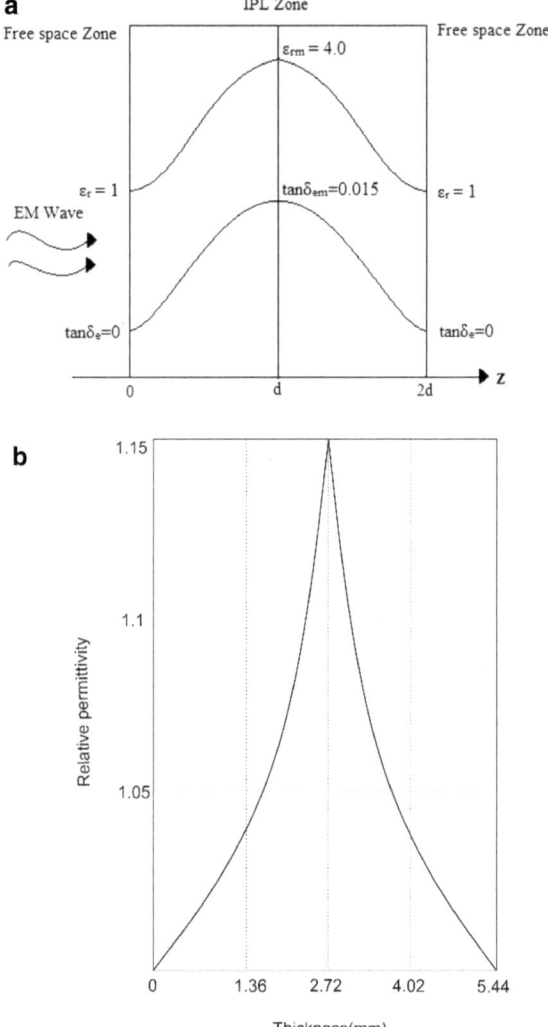

A-sandwich radome with IPL core: Thickness of the skin (Glass epoxy; $\varepsilon_r = 4.0$ and $\tan\delta_e = 0.015$) is 0.75 mm. The thickness of the core (Foam; $\varepsilon_r = 1.15$ and $\tan\delta_e = 0.002$) is 5.44 mm. The exponential variation of relative permittivity across the A-sandwich radome is shown in Fig. 6.1b.

Exponentially varying dielectric constant across the wall is given by (Philippe 1973),

$$\varepsilon_r(z) = \chi + \left(\frac{b^2}{[a(d-z)+1]^2} \right) \quad \text{for } 0 \le z \le d \tag{6.1}$$

$$\varepsilon_r(z) = \chi + \left(\frac{b^2}{[a(z-d)+1]^2} \right) \quad \text{for } d \leq z \leq 2d \tag{6.2}$$

In this section, the curvature parameter $\chi = 0.99$, d is half of the radome wall thickness, and the maximum value of dielectric constant is ε_{rm}. In case of monolithic IPL radome, ε_{rm} reaches the maximum value of 4 at its mid-plane as shown in Fig. 6.1a, similarly in case of A-sandwich radome the core dielectric constant ε_{rm} reaches the maximum value of 1.15 at its core mid-plane.

The coefficients, a and b, are given by

$$a = \frac{\{-1 + [(\varepsilon_{rm} - \chi)/(1 - \chi)]^{1/2}\}}{d} \tag{6.3}$$

$$b = (\varepsilon_{rm} - \chi)^{1/2} \tag{6.4}$$

6.2 EM Performance Analysis of Monolithic IPL Radome and A-sandwich IPL Radome

Polynomial curve fitting is used to incorporate the exponential variation of dielectric loss tangent ($\tan\delta_e$) in the IPL. Here, $\tan\delta_e$ varies from 0 and attains a maximum value of $\tan\delta_{em}$ at the mid-plane of the wall (in this case, $\tan\delta_{em} = 0.015$ as shown in Fig. 6.1a). Equivalent transmission line model (Chen et al. 2010) is used to compute the EM performance parameters of IPL radome. It is assumed that the IPL radome wall is made up of thin laminations (total number of laminations, $n = 744$ and each lamination thickness $= 0.01$ mm). The entire IPL radome wall represented by the cascaded voltage–current transmission matrix is given as

$$\begin{bmatrix} A & B \\ C & D \end{bmatrix} = \begin{bmatrix} \cos\Phi_1 & j\frac{z_1}{z_0}\sin\Phi_1 \\ j\frac{z_0}{z_1}\sin\Phi_1 & \cos\Phi_1 \end{bmatrix} \begin{bmatrix} \cos\Phi_2 & j\frac{z_2}{z_0}\sin\Phi_2 \\ j\frac{z_0}{z_2}\sin\Phi_2 & \cos\Phi_2 \end{bmatrix}$$
$$\begin{bmatrix} \cos\Phi_n & j\frac{z_n}{z_0}\sin\Phi_n \\ j\frac{z_0}{z_n}\sin\Phi_n & \cos\Phi_n \end{bmatrix} \tag{6.5}$$

Here, the electrical length corresponding to ith lamination is a function of the corresponding complex permittivity, the angle of incidence, wavelength λ, and the lamination thickness d_i where $i = 1, 2, 3, \dots n$.

The equations for voltage transmission coefficient and voltage reflection coefficient are given by (2.6) and (2.7).

The power transmission and power reflection coefficients are given by (3.9) and (3.10).

The IPD of IPL structure is given by

$$\text{IPD} = -(\angle T_1 + \angle T_2 + \cdots + \angle T_n)$$
$$-\frac{2\pi}{\lambda}(d_1\cos\theta_1 + d_2\cos\theta_2 + \cdots + d_n\cos\theta_n) \tag{6.6}$$

Here $\angle T_1$, $\angle T_2$, ..., $\angle T_n$ are the phase angles associated with the voltage transmission coefficients of the laminations with thicknesses d_1, d_2, d_3, ...,d_n, respectively. $\theta_1, \theta_2, \theta_3$, ..., θ_n are the corresponding incidence angles at the boundaries of laminations.

6.3 Numerical Results and Discussion for IPL Radome

For a conventional monolithic half-wave radome and IPL radome, the EM performance parameters are computed at normal incidence and at high incidence angles of 45° and 60° (Figs. 6.2a through 6.4c). EM performance degradation for perpendicular polarization exhibits the worst-case scenario, because the wave impedance for perpendicular polarization is higher as compared to that of parallel polarization. Therefore, EM performance parameters for perpendicular polarization alone are analyzed in this work.

At all incidence angles, the power transmission efficiency of IPL radome is superior to that of monolithic radome (Fig. 6.2a, 6.3a, and Fig. 6.4a). It is observed that the transmission efficiency of IPL radome is well above 90% throughout the 2–18 GHz frequency range for normal incidence, and for 45°and 60° incidence angles, the power transmission slightly degrades below 90% for frequency beyond 12 GHz.

The power reflection of IPL radome is well below 10% at all incidence angles in the entire frequency range (Fig. 6.2b, 6.3b and 6.4b). Low power reflection characteristics are desirable for minimization of the side lobe level (SLL) degradations of antenna radiation pattern and elimination of flash lobes in the radome applications.

The variation of IPD over the 2–18 GHz frequency range for monolithic IPL radome is less than that of monolithic radome, which indicates low phase distortions (Fig. 6.2c, 6.3c and 6.4c). This accounts for less boresight error (BSE), which is enviable for streamlined nosecone radome applications.

6.4 Numerical Results and Discussion for A-sandwich Radome with IPL Core

The power transmission performance curves of A-sandwich radome with IPL core having exponential variation of dielectric parameters for varying angles of incidence are shown in Fig. 6.5a through 6.7c. The performance characteristics of A-sandwich radome with IPL core show significant improvement at normal incidence, 45° and 60° than the A-sandwich alone.

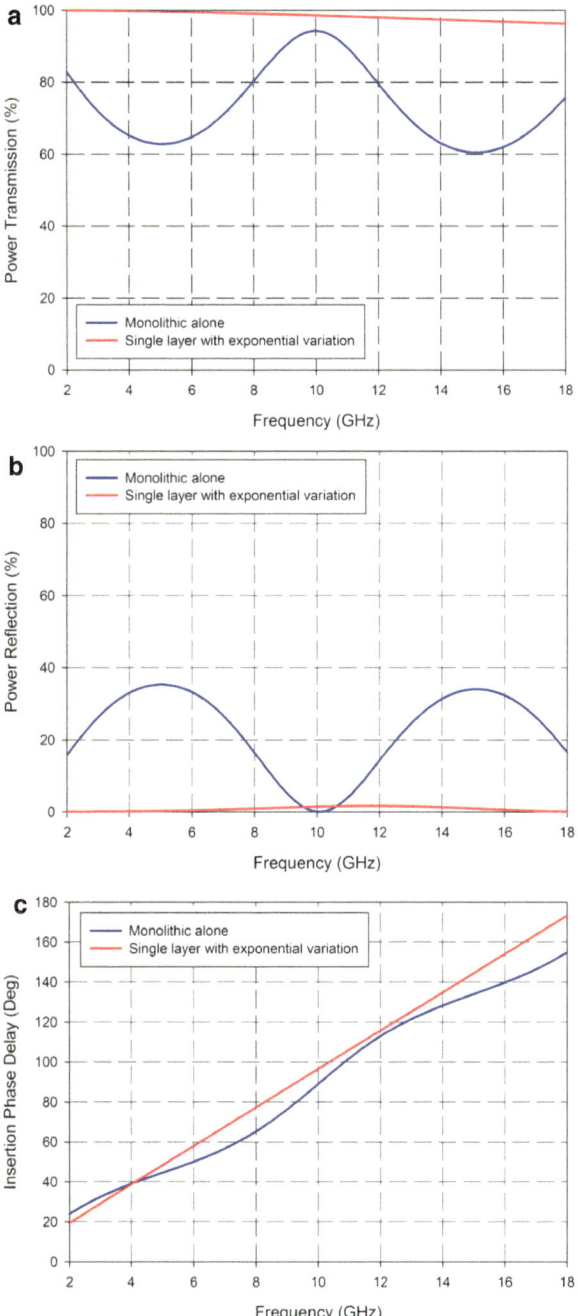

Fig. 6.2 a Transmission efficiency of IPL radome and monolithic radome at normal incidence.
b Reflection characteristics of IPL radome and monolithic radome at normal incidence. **c** Insertion
phase delay of IPL radome and monolithic radome at normal incidence

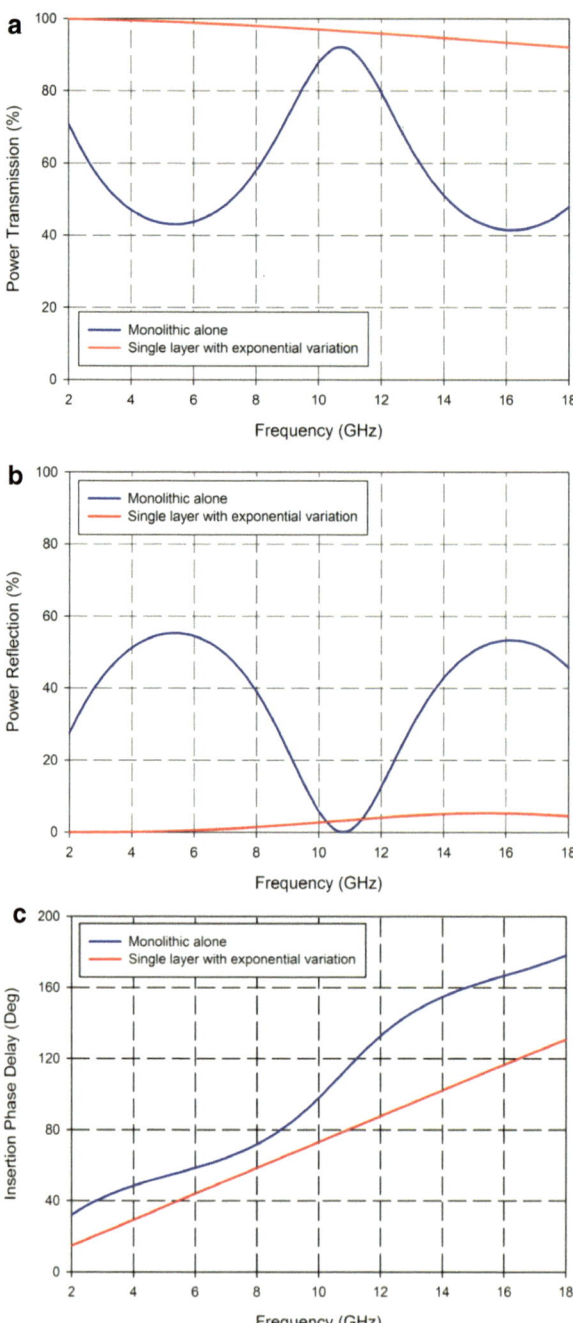

Fig. 6.3 a Transmission efficiency of IPL radome and monolithic radome at 45°. **b** Reflection characteristics of IPL radome and monolithic radome 45°. **c** Insertion phase delay of IPL radome and monolithic radome at 45°

Fig. 6.4 a Transmission efficiency of IPL radome and monolithic radome at 60°. **b** Reflection characteristics of IPL radome and monolithic radome at 60°. **c** Insertion phase delay of IPL radome and monolithic radome alone at 60°

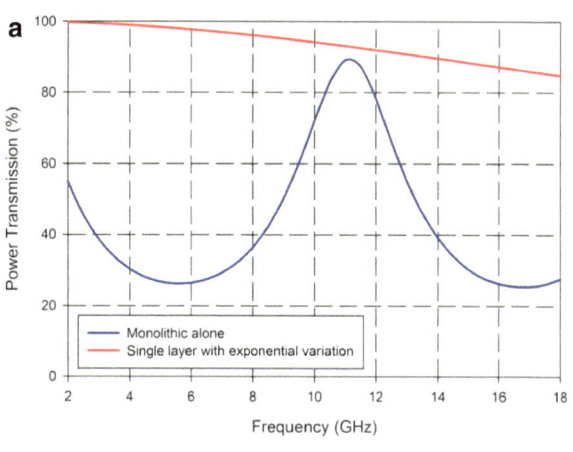

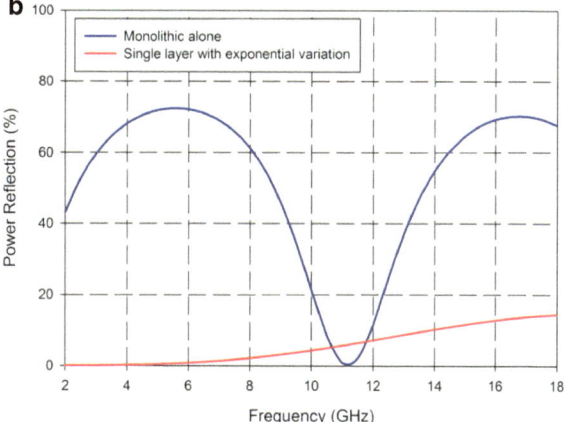

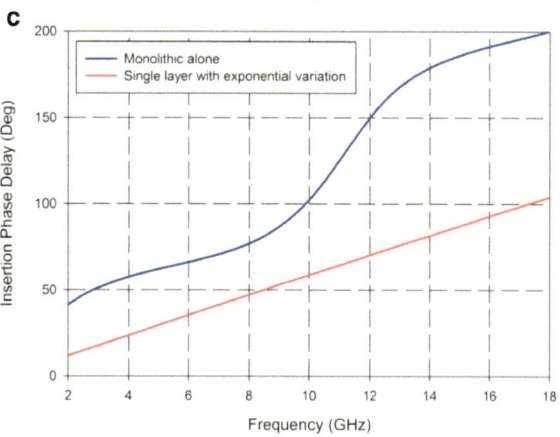

Fig. 6.5 **a** Transmission efficiency of A-sandwich radome with IPL core having exponential variation of complex permittivity and A-sandwich radome alone at normal incidence. **b** Reflection characteristics of A-sandwich radome with IPL core having exponential variation of complex permittivity and A-sandwich radome alone at normal incidence. **c** Insertion phase delay of A-sandwich radome with IPL core having exponential variation of complex permittivity and A-sandwich radome alone at normal incidence

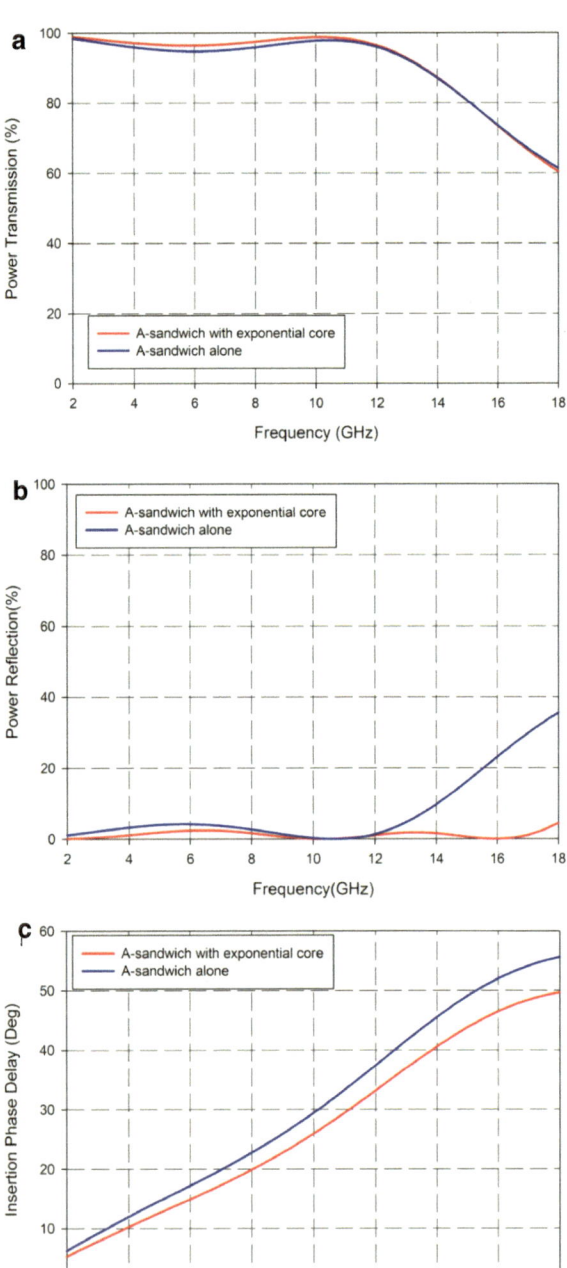

For normal incidence, the power transmission has a value greater than 95% till 12 GHz, beyond which it decreases. For incidence angles of 45° and 60°, the power transmission efficiency gets reduced significantly as shown in Fig. 6.5a, 6.6a, and 6.7a.

The power reflection of A-sandwich radome with IPL core having exponential variation is well below 10% at all incidence angles in the entire frequency range (Fig. 6.5b, 6.6b, and 6.7b).

The IPD performance characteristics curves of A-sandwich radome with IPL core having exponential variation of dielectric parameters are shown in Fig. 6.5c, 6.6c, and 6.7c. The IPD of IPL structure is less than the IPD of A-sandwich alone at all the angles of incidence considered in the study.

6.5   Conclusion

The EM performance of A-sandwich alone and A-sandwich radome with IPL core has been discussed in this section. The performance at normal, 45° and 60° incidence angles has been analyzed. The monolithic IPL radome design proposed in this section has superior EM performance characteristics as compared to conventional monolithic radomes at the three incidence angles considered here. Thus, the study establishes that the novel IPL wall configuration presented in this work is a better choice for the design of radomes. It is noted that A-sandwich with exponential varying IPL core shows no significant variation in the performance as compared to A-sandwich alone.

Fig. 6.6 a Transmission efficiency of A-sandwich radome with IPL core having exponential variation of complex permittivity and A-sandwich radome alone at 45°. **b** Reflection characteristics of A-sandwich radome with IPL core having exponential variation of complex permittivity and A-sandwich radome alone at 45°. **c** Insertion phase delay of A-sandwich radome with IPL core having exponential variation of complex permittivity and A-sandwich radome alone at 45°

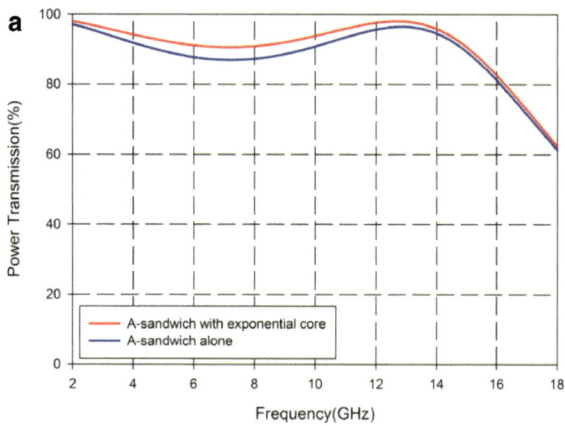

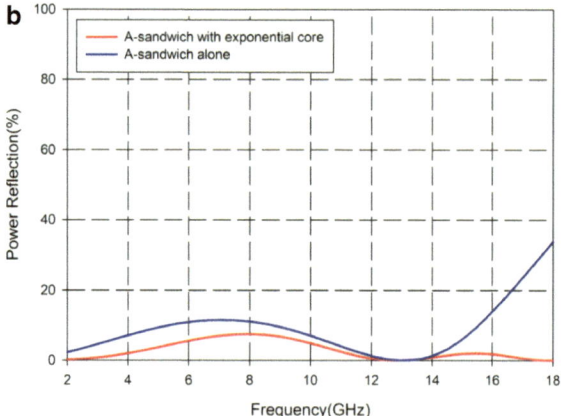

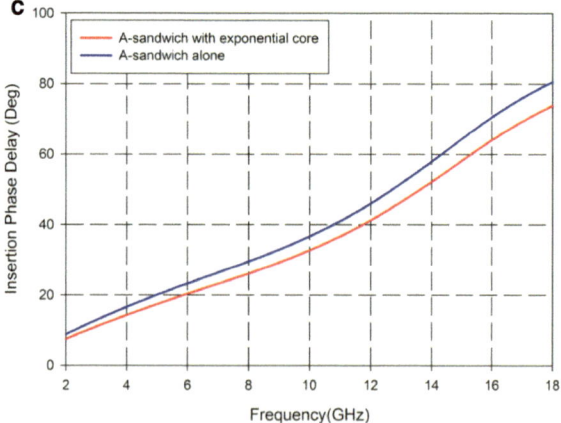

Fig. 6.7 a Transmission efficiency of A-sandwich radome with IPL core having exponential variation of complex permittivity and A-sandwich radome alone at 60°. **b** Reflection characteristics of A-sandwich radome with IPL core having exponential variation of complex permittivity and A-sandwich radome alone at 60°. **c** Insertion phase delay of A-sandwich radome with IPL core having exponential variation of complex permittivity and A-sandwich radome alone at 60°

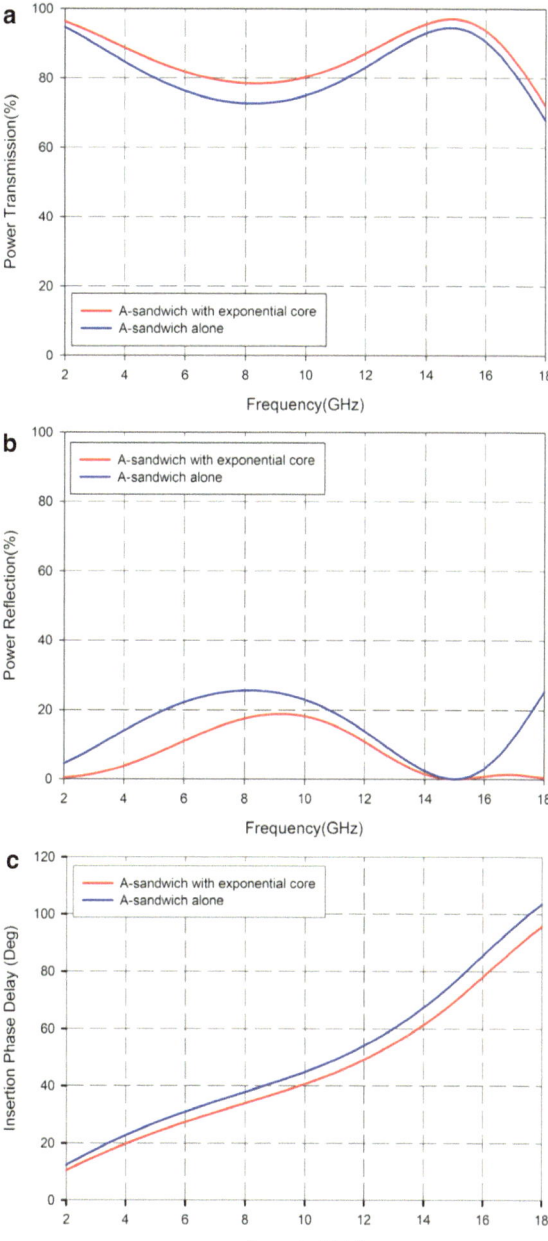

References

Amirhosseini MK (2006) Analysis of lossy inhomogeneous planar layers using Taylor's series expansion. IEEE Trans Antennas Propag 54(1):130–135

Chen F, Shen Q, Zhang L (2010) Electromagnetic optimal design and preparation of broadband ceramic radome material with graded porous structure. Progr Electromag Res 105:445–461

Philippe AE (1973) Reflection and transmission of radio waves at a dielectric slab with variable permittivity. IEEE Trans Antennas Propag AP-21(2):234–236

Toscano A, Vegni L, Bilotti F (2001) A new efficient method of analysis for inhomogeneous media shields and filters. IEEE Trans Electromagn Compat 43(3):394–399

Venkatarayalu NV, Ray T, Gan YB (2005) Multilayer dielectric filter design using a multiobjective evolutionary algorithm. IEEE Trans Antennas Propag 53(11):3625–3632

Summary

The present brief dealt with the design and analysis of radomes with inclusions embedded in wall of radomes for broadbanding purpose. From the analysis, it is evident that the presence of inclusions has an inevitable role in the performance enhancement of radomes. Such types of radomes give better performance as compared to conventional radomes.

The different techniques used to modify the monolithic and multilayered radome wall configurations are inclusion of metallic wire grids/strips in the radome wall, inclusion of metallic strip gratings in the radome wall, inclusion of frequency selective surfaces (FSS) in the radome wall, and use of inhomogeneous dielectric structures as radome wall.

The metallic inclusions used for broadbanding purpose can deteriorate EM performance of the radome as these structures may be sensitive to polarization and angle of incidence. Hence, the design parameters for the metallic structures have to be appropriately selected to obtain the desired performance.

In this brief, several techniques for broadbanding of radomes have been demonstrated. The analysis of the modified structures is done using the equivalent transmission line model. A detailed chapter-wise explanation for the design aspects and performance analysis of the modified radome wall configurations has been discussed in this book.

© The Author(s), under exclusive license to Springer Nature Singapore Pte Ltd. 2020   73
P. S. Mohammed Yazeen et al., *Broadbanding Techniques for Radomes*,
SpringerBriefs in Computational Electromagnetics,
https://doi.org/10.1007/978-981-33-4130-2

Author Index

Subject Index

© The Author(s), under exclusive license to Springer Nature Singapore Pte Ltd. 2020
P. S. Mohammed Yazeen et al., *Broadbanding Techniques for Radomes*,
SpringerBriefs in Computational Electromagnetics,
https://doi.org/10.1007/978-981-33-4130-2